CMOS Circuit Design –

Analog, Digital, IC Layout

CMOS Circuit Design

Analog, Digital, IC Layout

Nicholas L. Pappas. Ph.D.

ISBN-13: 978-1500569648   ISBN-10: 150056964X

A Message about this Text: The subject is essentially endless. The purpose here is to say enough about the subject so that you, the reader, have a running start when you apply this knowledge to your work.

Electronic circuit analysis, algebraic skills, and elementary digital logic design skills are essential prerequisites. Knowledge of some elementary calculus is also a prerequisite.[1]

We believe important benefits accrue by doing the problems carefully, by reconstructing the Spice programs and running them to reproduce the text figures, by deriving the text equations, and by drawing the layouts. These efforts provide "startup" work experience.

Once you have some work experience we are confident that you will be able to expand your know how with reasonable effort.

A Message from the Author: I have worked continuously in the electronics industry since 1950 except for 11 semesters teaching at San Jose State University (Professor and Chair Computer Engineering 1988-1993). There I discovered my talent for teaching such as it may be. After War2 I attended Lehigh University, and then transferred to Stanford where I earned the MS degree and, while working at HP in the early 1950's, the Ph.D. EE degree. (Somehow I did not get the word and formally apply for the BS degree.) Hardware design has been my principal activity. I learned enough about assembly language, Forth, C and C++ to design the software I needed for my projects. My current activity is designing integrated circuits.

---

[1] Thompson and Gardner, *Calculus Made Easy*,   ISBN 0312 185 480 (pbk)
Morris Kline, *Calculus*,   ISBN 0486 404 536 (pbk)

Preface

This text is different from other CMOS design texts, because not only do we actually show how to design CMOS circuits selecting transistor Length, Width and *the correct value of mobility* (a small detail that is usually overlooked if not ignored) we show how to make accurate, functioning circuit layouts that can be used in a chip. Furthermore we ask you to work hard drawing over 60 layouts using LASI that give you real world experience.

> *This is not about logic design.* This is about designing circuits and especially about making layouts suitable for integrated circuits. This is about implementing well known, widely used, digital and analog circuits as CMOS circuits.

CMOS circuits are the preferred technology for implementing modern digital integrated circuits. Why? A major economic reason is the relatively low cost of producing MOS devices and circuits, because of the extensive CMOS fabrication infrastructure investment that makes low cost possible. Two major technical reasons are (1) no DC current flows, (2) pmos and nmos transistors with reduced channel length L have significantly reduced on-off switching time resulting in faster and faster circuits.

We start by showing, step by step, how layouts are made that conform to Mosis[1] rules. We believe our presentation of the zillion details will allow you to absorb the details as you proceed so that you can layout circuits in parallel with the design presentations. Then you will be *able to do.*

A brief review of MOS transistors sets the stage for CMOS circuit design, which includes design of the MOS transistors used in the circuit.

Digital circuits with no memory implement logic equations as sums of minterms (OR of ANDs) or products of maxterms (AND of ORs). We show how to design circuits such as NOT (Inverter), NAND, NOR, XOR, Multiplexer, and Adder. Circuits are derived from their equations in a straightforward way. Spice programs evaluate performance. As we proceed we show how to plan and execute layouts for each circuit, which we ask you to layout as a parallel activity.

---

[1] www.mosis.com

CMOS Circuit Design

One bit digital circuits with memory are used in state machines. The RS Latch is the most elementary one-bit circuit with memory. Latches do not have clock inputs, whereas flip-flops and edge triggered flip-flops are one-bit memory circuits with clock inputs. The flip-flops are synchronous circuits. The D and JK edge triggered flip-flops are the flip-flop circuits in commercial use today.

Next the emphasis is on digital circuits that are an assembly of identical cells, such as the cell of a shift register. The integrated circuit layout of an assembly of cells is an orderly, repetitive pattern. Orderly, repetitive patterns are intrinsically free of layout errors. We say orderly layouts are mandatory for non trivial circuits (random logic layouts are high risk). We show how we make orderly layouts, and write Spice programs to evaluate performance.

We design and layout well known digital circuits such as shift registers, storage registers with load control, registers on a bus, and programmable logic arrays of logic with no memory.

The well known current mirror, differential amplifier, operational amplifier, resistors and capacitors are designed and their performance is evaluated by Spice. Layout procedures for the circuits as well as resistors and capacitors are presented.

Spice is used to plot DC response, AC frequency response, and TRAN transient response performance of circuits that are analyzed and designed in the text. You see results immediately. We show how to write these programs.

We ask you to draw layouts, which we consider to be useful *experiments* that give you real world experience. We consider drawing the more than 60 layouts to be a significant learning activity.

Chapter abstracts are next.

> Our blog *npappasee.blogspot.com* may offer you additional information. Take a look.

> We would appreciate receiving your comments and views on this text at *npappasz@yahoo.com.*

1 MOS Circuit Layout Most CMOS circuits are assemblies of pmos transistors, nmos transistors, and wires. We show every step when we show how to layout circuits using Mosis design rules. Layouts are used to build chips. *Realistic circuit layout procedures are used.*

LASI is the name of the widely used software used here. We show how to download and set up LASI.

The important topic of electromigration is discussed. Fast circuits require larger currents that emphasize the need to know about and understand electromigration, which can create open circuits in wires.

2 MOS nmos and pmos Transistors CMOS circuits are essentially assemblies of pairs of pmos and nmos transistors. Design equations for transistor current show the dependence on channel length L, width W, and mobility μ. Spice is used to plot the non-linear voltage current *vi* constraint for specific values of L, W, and μ.

There are two sets of design equations for MOS transistors referred to as long channel and short channel equations based on the electric field magnitude in the channel.

Transistor current is essentially proportional to mobility. Mobility plots allow for selection of the correct value for a specific transistor design.

3 Digital Circuits with no Memory Digital circuits with no memory implement logic equations as sums of minterms (OR of ANDs) or product of maxterms (AND of ORs).

We show how to design NOT, NAND, NOR, XOR, multiplexer, and full adder carry circuits. Circuits are derived from their equations in a straightforward way. Spice programs evaluate performance. As we proceed we show how to plan and execute layouts of these circuits.

Traditionally circuit layouts are placed in cells whose dimensions are defined by a frame so that arrays of cells can fit neatly side by side in rows. In turn the rows can arranged as rows one above the other. A frame template is defined for this purpose.

4 One Bit Digital Circuits with Memory Circuits with memory are used in *sequential circuits*. We prefer the modern name *state machines*.

Memory is implemented by *feedback* of a circuit output to a circuit input. The RS Latch is the most elementary one-bit circuit with memory. Latches do not have clock inputs, whereas flip-flops and edge triggered flip-flops are one-bit memory circuits with clock inputs. These flip-flops are *synchronous* circuits. The D and JK edge triggered flip-flops are the flip-flop circuits commercially available today.

Truth tables are derived from the defining equations. *A flip-flop's next state defining equation is the most important fact a designer needs to know.* The ideas of *Setup* and *Hold Windows* show how to guarantee correct *synchronous* circuit operation.

5 Complex Digital Circuits with Memory Next the emphasis is on circuits that are an assembly of cells, such as a 2 bit shift register cell. The integrated circuit layout of an assembly of cells is an orderly, repetitive pattern. Orderly, repetitive patterns are intrinsically free of layout errors. We say orderly layouts are mandatory for non trivial circuits (random logic layouts are high risk). We show how to make orderly, systematic, layouts, and write Spice programs to evaluate performance.

We design and layout well known circuits such as shift registers, storage registers with load control, registers on a bus, and programmable logic arrays implementing logic equations.

6. Analog Circuits Two widely used analog circuits, the current mirror, and differential amplifier are presented and analyzed. They are the basic building blocks for many analog integrated circuits.

The effect of channel length L and width W on MOS output resistance, bandwidth and gain is evaluated by Spice programs in the current mirror and differential amplifier designs.

Then a differential amplifier, an operational amplifier, resistors, and capacitors are designed, their performance is evaluated by Spice and layouts are drawn.

A key MOS design decision is selection of channel width W and length L for each MOS transistor in a circuit using *the correct value of mobility* μ. We show how to select values for W, L and μ.

7 CMOS Circuit Design Fundamentals Parasaitic capacitance increases on/off switching time and input to output signal delays. We show how circuit propagation delay depends on circuit C_{LOAD}. The capacitor C_{LOAD} represents the physical wires and MOS gates connected to a circuit output. The dependence of circuit delay on input signal switching time is analyzed.

Fanout is the maximum number of circuit inputs any circuit output is designed to drive. We derive an equation relating fanout, rail to rail switching time, and upper clock limit.

Signal delay on wires is analyzed as a function of wire resistance and capacitance.

8 How to write AC, DC, and TRAN Spice Programs Learning the Spice language per se is a very big project. However, knowledge of about 20 key words allows you to write the programs in this text. We show how to use those key words.

Spice is used to plot performance of DC response, AC frequency response, and TRAN transient response of circuits analyzed and designed in the text. You see results immediately. We show how to write these programs.

We recognize that you do not have to crunch numbers today, because Spice is the modern way to crunch numbers. That is why we include Spice in the text, and do not delegate it to labs.

Spice has an important role in the modern design process.

Appendix
A1 Download and install Lasi
A2 The files 180_N1P1.txt, mos018_2.lib and mos018_L.lib

Contents

Universal Constants and Other Symbols

$$\varepsilon_0 = 8.855 \times 10^{-12} \frac{F}{m} \approx \frac{10^{-9}}{36\pi} \frac{F}{m} \qquad \textit{dielectric constant of free space}$$

$$\mu_0 = 4\pi 10^{-7} \frac{H}{m} \qquad \textit{permeability of free space}$$

$$c = \frac{1}{\sqrt{\mu_0 \varepsilon_0}} \approx 2.998 \times 10^8 \frac{m}{s} \qquad \textit{velocity of light}$$

$$z_0 = \sqrt{\frac{\mu_0}{\varepsilon_0}} \approx 120\pi \ ohms \qquad \textit{characteristic impedance of free space}$$

$$q_e = 1.602 \times 10^{-19} \frac{coulombs}{electron} \qquad \textit{charge of the electron}$$

$$k = 1.381 \times 10^{-23} \frac{J}{°K} \qquad \textit{Boltzman's constant}$$

$T \quad °K \qquad \textit{temperature, degrees Kelvin}$

$N_A \dfrac{carriers}{m^3} \qquad \textit{carrier concentration}$

$n_i \dfrac{carriers}{m^3} \qquad \textit{intrinsic carrier concentration}$

$\varepsilon \dfrac{F}{m} \qquad \textit{dielectric constant of the material}$

$$\varepsilon_{SiO_2} = 3.97\varepsilon_0$$

$$\varepsilon_{Si} = 11.9\varepsilon_0$$

The SI System of Units

SI Prefixes

Prefix	Multiplier	Symbol
tera	10^{12}	T
giga	10^{9}	G
mega	10^{6}	M
kilo	10^{3}	K
milli	10^{-3}	m
micro	10^{-6}	μ
nano	10^{-9}	n
pico	10^{-12}	p
femto	10^{-15}	f
atto	10^{-18}	a

Basic Units

Quantity	Name	Symbol
Length	meter	m
Mass	kilogram	kg
Time	second	s
Current	ampere	A
Temperature	Kelvin	K
Luminous Intensity	candela	cd

Derived Units

Quantity	Name	Formula	Symbol
Acceleration	meter per sec per sec	m/s^2	a
Velocity	meter per sec	m/s	v
Force	newton	$kg \times m/s$	N
Pressure (stress)	pascal	N/m^2	Pa
Density	kg per cubic meter	kg/m^3	r
Energy or work	joule	$N \times m$	J
Power	watt	J/s	W
Charge	coulomb	$A \times s$	C
Potential	volt	$W/A = J/C$	V
Resistance	ohm	V/A	Ω
Capacitance	farad	C/V	F
Magnetic flux	weber	$V \times s$	Wb
Inductance	henry	Wb/A	H

Greek Alphabet

A	α	alpha	a[1]
B	β	beta	b
Γ	γ	gamma	g
Δ	δ	delta	d
E	ε	epsilon	e
Z	ζ	zeta	z
H	η	eta	h
Θ	θ	theta	q
I	ι	iota	i
K	κ	kappa	k
Λ	λ	lambda	l
M	μ	mu	m
N	ν	nu	n
Ξ	ξ	xsi	x
O	o	omicron	o
Π	π	pi	p
P	ρ	rho	r
Σ	σ	sigma	s
T	τ	tau	t
Y	υ	upsilon	u
Φ	φ	phi	f
X	χ	chi	c
Ψ	ψ	psi	y
Ω	ω	omega	w

---

[1] equivalent computer keyboard English letter keys

1 MOS Circuit Layout

Most CMOS circuits are assemblies of pmos transistors, nmos transistors, and wires. Pmos transistors have channels that conduct holes, whereas nmos transistors have channels that conduct electrons.

Think of a p wafer surface (1.1 The Semiconductor Wafer and Wells) as a desert that stretches as far as the eye can see. This is the view you would have were you 10 micrometers tall. Don't ask us how, but nearby you discover two tubs W micrometers wide labeled S and D (Figure 101). The tubs are separated by a (W+2Δ) L area labeled G (*ignore the G extensions Δ beyond W*). The tubs have high conductivity, and are marked as n+.

Closer examination reveals the (W+2Δ)L area is covered with a sheet of pure 'glass' that is a non-conducting silicon dioxide material. There is more – covering the glass is a sheet of material with a label on it. The label says "this is conductive *polysilicon*" – whatever that is – so that when electrons flow through it only an acceptable voltage drop results. The polysilicon (poly) "wire" extends a fraction of a micrometer out into the desert beyond both ends of the n+ tubs that we ignore. You discover that if you apply a positive voltage to the polysilicon and ground the wafer, the W×L area under G fills with electrons.

Figure 101 MOS

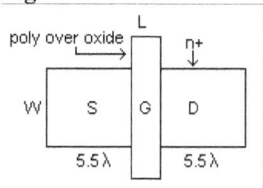

This area with dimensions W×L has another label on it which says "this is an *n channel* L micrometers long and W micrometers wide." You have discovered an nmos transistor in the p desert! (If this was an n wafer you would have discovered a pmos transistor with S and D p+ tubs.)

The basic idea The MOS transistor is controlled by the basic law Q=CV about capacitors C. A positive voltage across the terminals of capacitor C results in a charge +Q on the capacitor plate that is the poly over G (Figure 101), and an equal and opposite charge –Q (electrons) on the other plate that is the p wafer surface under G. Since Q=CV, a change in voltage induces a change in the quantity of electrons available for flow in the *n channel* from S to D. The transistor channel is the W×L area under the *gate oxide*.

In other words the *SiO₂ gate oxide* is the MOS capacitor dielectric that is between the plates. One conducting plate is poly or metal (in the pre-poly era) which is plated on the top side of the SiO₂ layer. This is the gate terminal. The other conducting plate is the silicon surface under the SiO₂, which was grown as a sheet. This is the C_{gate}. The area under the gate oxide becomes the channel where current flows from S to D in the L direction. The electric field in the gate oxide provides the *field effect* of the gate voltage on the channel current. $C_{gate}=C_{ox}\times W$ micrometers $\times L$ micrometers where C_{ox} is the gate capacitance per square micrometer.

$$(1) \quad C_{gate} = C_{ox}WL = \frac{\varepsilon_{SiO2}\varepsilon_0}{t_{ox}}WL = \frac{3.9 \cdot 8.85 \cdot 10^{-12}}{4 \cdot 10^{-9}}W_{\mu m}L_{\mu m}10^{-12}$$

$$= 8.63W_{\mu m}L_{\mu m}10^{-15} \ farads$$

The n+ regions on either side of the channel (Figure 101) are connected to drain D and source S terminals. The n+ regions are also known as *active* regions. Active is a generic name for n+ or p+ regions. The MOS transistor has four terminals for circuit connections. They are drain n+, gate poly, source n+, and the body, which is the wafer. A positive voltage on the gate induces negative charge in the gate capacitor's p silicon surface plate so that a drain to source voltage V_{DS} can use these electrons to create a drain current I_{DS} in the channel. The field effect is why the gate voltage V_{GS} controls the magnitude of the channel current. In other words a conducting channel from source to drain is created by using the charge induced in the channel by a voltage connected from gate to source.

Constructing a transistor Semiconductor wafers can be n wafers with an excess of electrons, or p wafers with an excess of holes. In practice p wafers are used. Since nmos transistors are fabricated in p wells you can place nmos transistors at any location in a p wafer that is just one p well. On the other hand pmos transistors are fabricated in n wells. Consequently you form wells of n regions anywhere in the p wafer so that you can place pmos transistors in those n wells. We only discuss the essentials of the CMOS fabrication process, because this text is not about fabrication per se. We show how to make transistor layout drawings that satisfy Mosis[1] design rules. If you satisfy a set of design rules, then your layout can be used to produce a chip.

Electromigration Fast circuits require larger currents that emphasize the need to know about and understand electromigration, which can create open circuits in wires.

---

[1] www.mosis.com

1.1 The Semiconductor Wafer and Wells

Solid state physics is essentially the study of crystals, and behavior of electrons in those crystals. A wafer is a properly cut slice of a silicon crystal lattice. The following discussion is conceptual, and not rigorous.

Electrons in the outer shell A silicon atom has four electrons in its outer electron shell that can hold 8 electrons. This means silicon atoms can form a crystal which has a diamond like lattice, because each atom is bonded to four adjacent silicon atoms by valence forces established by four outer shell electron pairs. Each electron pair consists of one electron from each of two silicon atoms. The four electrons in atom #1 are bonded to one electron each in atoms #2, #3, #4, and #5. The four electrons in atom #2 are bonded to one electron each in atoms #1, #6, #7, and #8, and so forth so that all atoms have 8 outer shell electrons .

The silicon crystal is an insulator at zero degrees Kelvin. At higher temperatures some electrons escape from their atomic shells to form an electron gas in thermal equilibrium with the atomic lattice. Consequently the silicon crystal has an electric conductance in the range 10^{-6} to 10^{2} Siemens (formerly mhos) that is less than the 10^{7} mhos electric conductance of metals and greater than the 10^{-10} mhos electric conductance of insulators. The silicon crystal is a *semiconductor*; it is not a conductor nor is it an insulator. The silicon crystal is charge neutral.

Donors Consider arsenic atoms which have five electrons in their outer electron shell. If relatively few arsenic atoms are diffused into the pure silicon wafer, then those arsenic atoms displace silicon atoms, one for one, in the crystal lattice. Four of the five arsenic outer electrons bond with one electron in each of the four adjacent silicon atoms. The fifth electron, call it a −1 charge, is not bonded. The unbonded (arsenic) electron is free to roam in the lattice. When an electron leaves an arsenic atom the atomic charge that changes from 0 to +1 is not mobile. Therefore a surface enclosing the wafer has a net zero charge inside. The wafer is said to have been doped to become an n wafer that can conduct "the 5th" electrons donated by the arsenic atoms. The conduction electrons are referred to as donors.

Acceptors Next consider boron atoms which have three electrons in their outer electron shell. If relatively few boron atoms are diffused into the pure silicon wafer, then those boron atoms displace silicon atoms, one for one, in the crystal lattice. The three boron outer electrons bond with one electron in each of three adjacent silicon atoms. The fourth adjacent silicon atom has an electron, a −1 charge that is not bonded. Since one bond is unsatisfied the world says a hole *can be* created. An unbonded (silicon) electron is free to roam in the lattice. When an electron leaves a silicon atom a hole is created, and the silicon atom charge changes from −1 to +1. The electron roams until it is captured by a +1 charge and fills some other hole. This is how holes appear and disappear. The holes roam in the lattice. The wafer is said to have been doped to become a p wafer which can conduct holes. The conducting holes are referred to as acceptors.

P wafer is one p well (boron doped) A negative potential over a surface area creates an electric field that *repels* any shaken-loose electrons from the volume just below the surface area uncovering the unsatisfied bonds of the silicon atoms created by dopant boron atoms thereby resulting in a surplus of holes in the volume just below the surface. No electrons are available for conduction. If the surface potential is changed from negative to zero, then the below-surface volume returns to the zero charge state. Continuing the change, the surface potential becomes positive and shaken-loose electrons are attracted to the volume just below the surface to fill holes. The accumulation of electrons is interpreted as a depletion of holes. If the surface potential continues to become more positive yet more electrons are attracted to the below-surface volume until the density of electrons exceeds the density of holes. The below-surface volume experiences inversion from positive charge density to negative charge density. Free electrons are now available to become conduction electrons when a suitable potential is applied. The n-channel that has been created forms the basis for the nmos transistor. In other words a *threshold voltage* V_{TN} created the n channel.

N wells in a p wafer (arsenic doped) An n well is a volume of the p wafer into which arsenic atoms have been diffused to some depth, in sufficient numbers, so as to fill the p wafer holes and have a surplus of free electrons. In other words the n well volume is now a partial n wafer in which we can fabricate pmos transistors.

A positive potential over a surface area in the n well creates an electric field that attracts free electrons from the volume below the surface area thereby resulting in an accumulation of electrons in the volume just below the surface. No holes are available for conduction. If the surface potential is changed from positive to zero, then the below-surface volume returns to the zero charge state. Continuing the change the surface potential becomes negative and the free electrons are repelled from the volume just below the surface to uncover the fixed positive charge of the arsenic atoms. The removal of mobile electrons is a depletion of electrons. If the surface potential continues to become more negative yet more electrons are repelled from the below-surface volume until the density of positive charge from uncovered arsenic atoms exceeds the density of free electrons. The below-surface volume experiences inversion from negative charge density to positive charge density. This excess of positive charge to negative charge density is interpreted as a density of holes. Holes are now available to conduct current when a suitable potential is applied. A p-channel has been created by a *threshold voltage* V_{TP} that forms a pmos transistor.

1.2 The MOS Fabrication Process

The CMOS fabrication process has many steps we *do not present* in a detailed sequence. We focus on layout essentials. Basic and complete circuit layout procedures are presented in upcoming paragraphs. CMOS technology implements MOS transistor circuits in a semiconductor wafer by using three types of conducting material: metal, polysilicon, and active (the n+ and p+ diffusions in Figures 101, 102 page 6).

> The metal, polysilicon, and active *conducting* layers are always separated from each other by an insulating oxide layer. Layer to layer connections are made by vias (holes) cut through the oxide.

In a layout, when a *polysilicon path crosses over an active area* an MOS transistor results (1.3 Drawing Layouts). Metal passing over polysilicon or active has essentially no circuit effect. Know that there may be more than one layer of metal. Metal filled contact holes (vias) through an insulating layer provide for connection of paths on different layers that cross over each other, such as metal path over polysilicon path, metal path over active area, and metal path over metal path (Figure 102). Polysilicon is *never* connected by contacts to active.

The metal and polysilicon layers have a multitude of paths representing wires as dictated by the circuit being implemented. (There may be 5 or more metal layers.) The polysilicon layer also has areas representing transistor gates. The active layer has areas for each MOS transistor drain and source, as well as for connections by contacts to the wafer surface.

Paths and areas for the three types of conducting layers and the intervening contact holes are transferred from drawings to photographic negatives known as masks. In turn, the masks are used during the fabrication process to form the patterns of paths, areas, and contact holes on each layer. In this way various circuits assembled from metal wires, polysilicon wires, and transistors become systems on chips cut out of a wafer.

Figure 102 Wafer cross-section showing nmos and pmos transistors

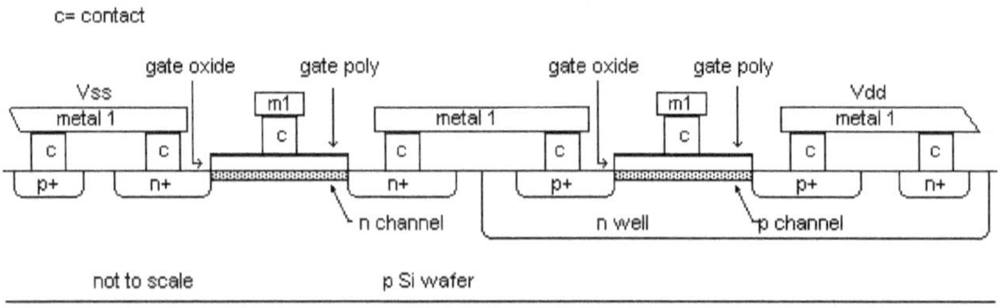

c= contact

gate oxide   gate poly   gate oxide   gate poly

Vss   Vdd

metal 1   m1   metal 1   m1   metal 1

c   c   c   c   c   c   c   c

p+   n+   n+   p+   p+   n+

n channel   n well   p channel

not to scale   p Si wafer

1.3 Drawing Layouts

Layouts are top views of wires and transistors. In what follows layout dimensions are dictated by *design rules* we use as a layout proceeds. *Realistic circuit layout procedures used* in upcoming paragraphs illustrate the process.

An MOS transistor cross-section (Figure 103) shows the polysilicon gate is over the gate oxide of thickness t_{ox} that, in turn, is over the silicon surface (p wafer). The active (n+ diffusion) regions at either end of the channel complete the transistor. (None of these figures are drawn

Figure 103 NMOS cross-section

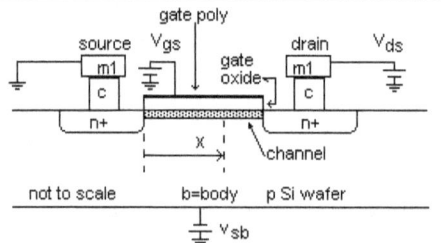

gate poly

source V_{gs} drain V_{ds}

     m1   gate   m1

     c   oxide   c

n+   n+

X   channel

not to scale   b=body   p Si wafer

V_{sb}

to scale.) The gate capacitor, C_{gate}, is formed by the polysilicon and channel plates separated by an insulating oxide t_{ox} thick. This is an example of polysilicon passing over an oxide/silicon area forms an MOS transistor. Observe that *there is no n+ active beneath the polysilicon*: there is only the n channel formed in the p surface.

Parasitic diodes The n+ diffusions/p-wafer and the n-channel/p-wafer junctions form unwanted pn diodes (Figure 103). The p+ diffusions/n-well and the p-channel/n-well junctions also form unwanted pn diodes. The n channel diodes are turned off by connecting the p wafer surface to V_{SB} volts (e.g. 0 volts) and energizing the nmos transistors with positive V_{DS} voltages. The p channel diodes are turned off by connecting the nwell surface to V_{DD} volts (e.g. 1.8 volts), and the pmos transistors are energized with positive voltages less than V_{DD}.

Integrated circuit layouts In order to make an integrated circuit layout you start with layout software, and a set of design rules for the process technology. Then you select the circuit you want to layout.

Lambda (λ) The unit dimension of a layout is one lambda (λ). Design rules say the smallest allowable channel length L is 2λ. When an improved process with smaller feature dimensions comes into production we simply change the value of lambda that automatically scales the drawings so that no modifications are necessary. However, sometimes we have to modify the drawings, because some design rules may change when lambda changes.

Draw a transistor The n+ and p+ diffusion areas are drawn as *active* areas. For example a channel is formed by drawing a $2\lambda \times 11\lambda$ *poly box* over a $7\lambda \times 13\lambda$ active area (Figure 107). Contacts ($2\lambda \times 2\lambda$ vias) to source and drain are implemented by adding (black) *contact* and (cross hatched) $4\lambda \times 4\lambda$ *metal$_1$* boxes creating an active/contact/metal three layer sandwich structure. Box dimensions are specified by design rules. Layer thickness is set by the semiconductor fabrication process. Note: The software erases the active area under the gate exposing the p wafer to the oxide under the poly. The oxide layer is added by the software.

Figure 107 Boxes defining a transistor with W/L=7λ/2λ (dots 1λ)

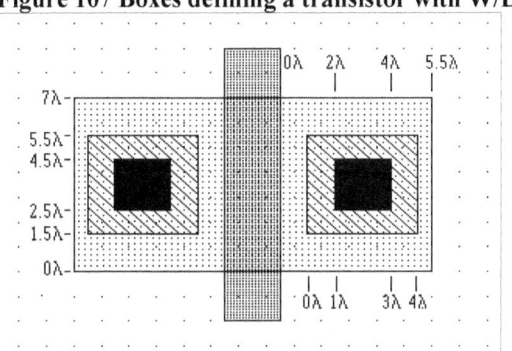

CMOS Circuit Design

An active area is defined as n or p when it is enclosed by a *n or p select box*. Pmos transistors must be placed in an n well (Figure 108).

Figure 108 n well and p-select boxes defining a pmos transistor

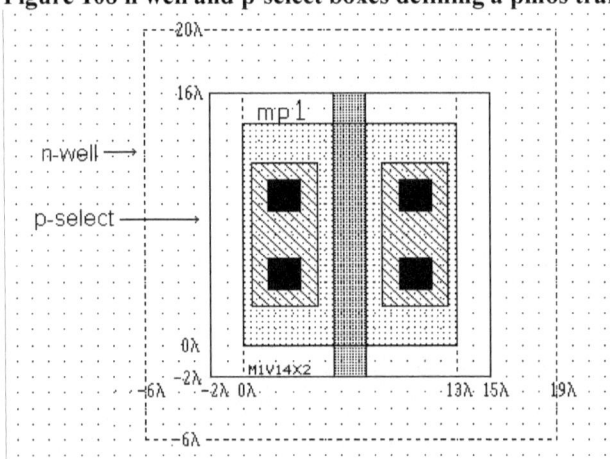

Important The minimum 5.5λ length of an active area (Figure 107) on either side of the poly gate is defined by contact rules 6.1, 6.2, 6.4 (2λ+1.5λ+2λ). Thus the minimum length of a transistor's active area is 13λ (5.5λ+2λ+5.5λ) as shown in Figure 108. Then the minimum active area is 13λ×W.

Rule	Simple Contact to Active	Lambda
6.1	Exact contact size	2x2
6.2	Minimum active overlap	1.5
6.3	Minimum contact spacing	3
6.4	Min spacing to gate of transistor	2

Metal layers wire a circuit Connections to transistor terminals are made by *contacts* (vias) connecting *active* to *metal₁*. In turn the metal₁ layer is connected to the metal₂ layer by a contact referred to as via₁. Vias connect any two conducting layers. We draw vertical wires in even numbered metal layers, and horizontal wires in odd numbered metal layers,

metal₁/via₁/metal₂
metal₂/via₂/metal₃
metal₃/via₃/metal₄ and so forth

8

Transistors in series and in parallel Transistors in series and in parallel constantly appear in circuits (Figure 104). You have the option to merge layouts of series and parallel pairs to save layout area. Merging layout boxes means placing one box over another. The drawing data base only retains one box (the logical AND).

Figure 104 MOS transistors in series and in parallel

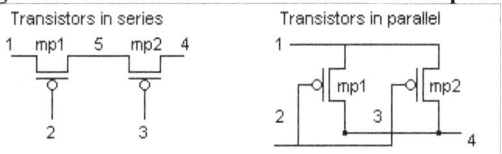

Transistors in series can be laid out with a wire from mp_1's drain to mp_2's source as drawn in the schematic (Figure 105a). Mp_1's drain can overlay mp_2's source, and contacts are retained makes node 5 available (Figure 105b). The node 5 contacts can be removed (Figure 105c). You can think of this layout as two poly paths crossing the active area forming two channels in a series connection. In any case current enters the series pair via two contacts, and exits via two contacts.

Figure 105 Layout of MOS transistors in series and in parallel

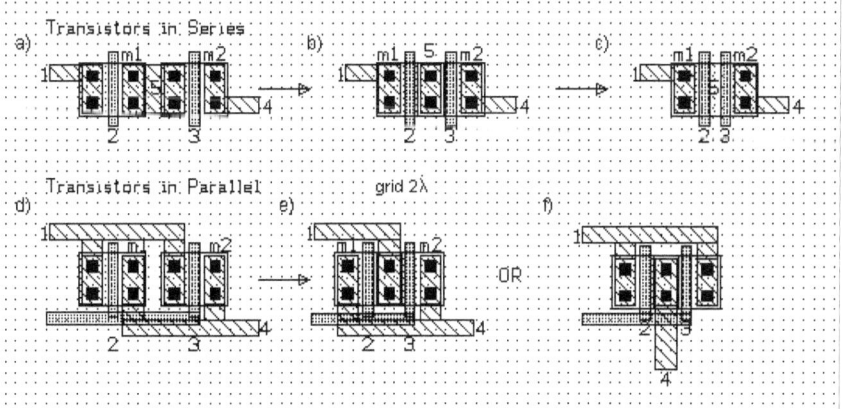

Parallel transistors are merged in the same way (Figure 105d, e, f). Merging is a straightforward way to make very wide transistors. Merge n transistors of width W to get an equivalent one of width n × W. Observe that you lose one set of contacts. The source has n sets and the drain has n−1 sets (or vice-versa).

Drawing cells Circuit design calculated the W's for the transistors in the circuit. One transistor is drawn for each W. Each transistor is in a cell of rank 1. A rank 2 circuit cell is created so that transistor rank 1 cells can be imported into the rank 2 circuit cell. Then metal wires on various layers wire the circuit.

The layout process In what follows process steps are marked as 1.1, 1.2, etc. These actions draw the required boxes. *Everything drawn is a box.*

Specific drawings are made in *cells*. Each cell is assigned a rank so that a hierarchy of cells can be developed. For example, individual transistors of various sizes are drawn in rank 1 cells. A NAND circuit in a rank 2 cell can incorporate rank 1 transistor cells. A logic circuit in a rank 3 cell can incorporate rank 2 NAND circuit cells, and so forth up the chain until the rank n cell containing the entire chip is reached.

Mosis Design rules The latest versions of Mosis scaleable CMOS design rules are available at

http://www.mosis.com

> The metal, polysilicon, and active *conducting* layers are always separated from each other by an insulating oxide layer. Layer to layer connections are made by vias (contacts) cut through the oxide.

Setup and Run the Drawing program A popular program is Lasi[2]. Lasi is a first class, reliable program. We use Lasi here. The setup process is spelled out in Appendix *A1 Download and install Lasi.*

We ask you to do the following as you read Chapter 1 so that at the end of the chapter you will have replicated all of the figures and will have acquired skill using Lasi.

> To fetch help while using the Lasi program
> hold down F1, while clicking on any button.

> Use Lasi to draw boxes as they are specified for each drawing.

---

[2] http://lasihomesite.com

1.3.1 Draw an MOS Transistor

One transistor with W/L=14λ/2λ is drawn as an original drawing according to Mosis design rules. A second transistor with W/L = 7λ/2λ is drawn by copying and modifying the first drawing. Sometimes we omit lambda λ when we write W/L ratios.

Lambda Feature sizes constantly decrease as time elapses. This is why design rules use normalized dimensions. The minimum basic length used in the drawings is set by the fabrication process. All dimensions are proportional to this fundamental unit of length referred to as λ (lambda). We use the layout software's one λ grid to avoid most decimal points when we specify coordinates.

Lasi Start the lasi program. All drawing activity is on the Lasi screen.

A transistor layout uses the three types of conducting layers: active, poly, and metal₁. The contact layer is an oxide. Contacts are *metal filled vias* in the contact oxide layer. Contacts connect metal to active or poly as required. Four drawing layers (active, poly, contact, and metal₁) are used to define an MOS transistor. Drawing a transistor requires the following steps.

1.1 Cell	1.2 Active	1.3 Poly
1.4 Contact	1.5 Metal₁	1.6/1.7 M1V07x2

1.1 Create Cell Define Cell M1V14x2. You need a rank 1 cell in which you will draw the transistor. (Use *Get* and *Del* to erase errors.)

•Click *Load (in upper tool bar)*, type in the cell name M1V14x2, and click OK. Rank dialog box appears. Type a 1. Click on OK. A screen with the origin marker and grid appears. Click on *Grid* to see the grid dots. Click on *Dgrd* until *Dgrd=1LAM* appears in the status line at screen bottom. Click on *Wgrd* to get *Wgrd=0.5LAM*. The characters in the name M1V14x2 represent specific items. The form of this name is our personal choice.

M	for mos transistor,
1	for one poly stripe,
V	for vertical poly stripe,
14	for channel width W in lambda units,
x	by
2	for channel length L in lambda units

1.2 Draw Active Box Active boxes are drawn in the ACTIVE layer that defines the ACTIVE mask. Contact rules define the active box 5.5λ length on both sides of poly. Transistor W defines the active box width.

Rule	Simple Contact to Active	Lambda
6.1	Exact contact size	2x2
6.2	Minimum active overlap	1.5
6.3	Minimum contact spacing	3
6.4	Min spacing to gate of transistor	2

Rule 6.4 specifies 2λ as the minimum distance from poly edge to contact edge. Rule 6.1 says contact width is 2λ. Rule 6.2 says active overlap is 1.5λ. So poly to active edge is $2+2+1.5$ units or 5.5λ is one source and drain dimension, and W is the other dimension (Figure 106). The active area length is $5.5\lambda+2\lambda+5.5\lambda =13\lambda$. The active area width is W. The area is $13\lambda\times W = 13\lambda\times14\lambda$ (Figure 106).

Figure 106 Transistor M1V14x2 W/L=14/2

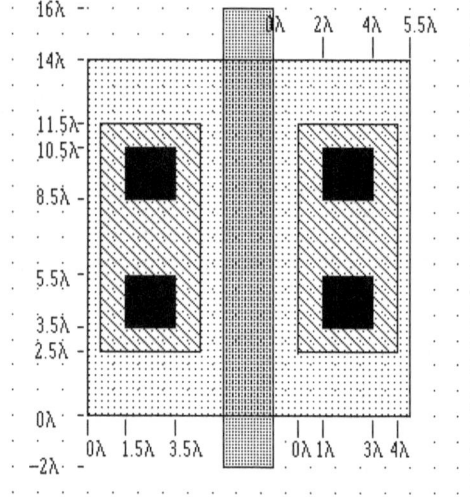

In *Menu 1* (Always watch the status bar)

•Click *List*. Double click cell M1V14x2 to see the cell origin.

•Click *Layr* and change the current layer to ACTV 43.

•Click *Obj*. Double click Box to change the current object to BOX.

•Click *Add*. Click on the U button. Click on the origin's grid dot at (0, 0), move the cursor to (13, 14) (the cursor snaps to the grid and note the "rubber band" box as the cursor moves). Click on a second grid dot with coordinates x=13, y=14. A 13×14 green box with green dot pattern appears.

Problem 101 Draw the Lasi layout for the transistor in Figure 106.

1.3 Draw Poly Box All poly boxes are drawn in the POLY layer.

Rule	Poly	Lambda
3.1	Minimum width	2
3.2	Minimum spacing over field	3
3.2a	Minimum spacing over active	3
3.3	Min gate extension of active	2
3.4	Min active extension of poly	3
3.5	Minimum field poly to active	1

Rule 3.1 specifies that minimum poly box dimension is 2λ (this is the MOS L of W/L). Rule 3.3 specifies minimum poly gate extension past any active edge must be 2λ. This is why we want a $2\lambda \times 18\lambda$ poly box, where 18λ equals 2 gate extensions of 2λ plus W of 14λ (Figure 106). W wide active defines the mos transistor gate width W. Poly length L defines mos transistor channel length L, which is set to 2λ for the fastest transistor.

•Click *Layr* and change the current layer to POL1 44.
•Click *Wgrd* to change the grid to 0.5λ spacing (watch the status bar).
•Click *Obj*. Double click Box to change the current object to BOX.
•Click *Add*. Click on the grid dot at (5.5, −2), move the cursor to (7.5, 16) and click to add a 2×18 poly box. A red box with red dot pattern appears.

1.4 Draw Contact Boxes The active and poly boxes define the basic transistor to which we now add contacts (Figure 106).

Rule 6.1 specifies contact size as $2\lambda \times 2\lambda$. Rule 6.3 specifies a minimum contact spacing of 3λ to the next box. We want as many contacts that will fit in W=14λ. Try 3 contacts. Three contact boxes require 1.5+2+3+2+3+2+1.5 or 15λ. W = 14λ so we can only use two contacts per column. Two contact boxes occupy 1.5+2+3+2+1.5 or 10λ minimum. Since W = 14λ the four remaining lambda are allocated to the top and bottom of the contact column to get 3.5+2+3+2+3.5 or 14λ (Figure 106).

•Click *Layr* and change the active layer to CONT 28.
•Click *Obj*. Double click Box to change the current object to BOX.

•Click *Add*. Add a 2λ×2λ contact box with (x,y) corners at (1.5, 3.5) and at (3.5, 5.5) (Figure 106). A blue box with solid fill appears. Verify that the right contact edge to poly edge is 2λ (Rule 6.4), and the left contact edge to active edge is 1.5λ (Rule 6.2).

•Click *Get*. Enclose the contact box with a rubber band box and click to make the box outline white. Always click *Draw* to redraw the drawing.

•Click *Cpy*. Click box corner (1.5, 3.5). Move cursor to box-to-be corner (1.5, 8.5) and click. A blue contact box appears

•Copy column of 2 contacts. Copy the column of boxes from corner (1.5, 3.5) to corner (9.5, 3.5).

•Click *Put*. Enclose the contact boxes with a rubber band box and click to erase the white lines.

1.5 Draw Metal₁ boxes

Rule	Metal$_1$	Lambda
7.1	Minimum width	3
7.2	Minimum spacing	3
7.3	Min overlap of any contact	1
7.4	Minimum spacing when either metal$_1$ line is wider than 10λ	6

Rule 7.3 specifies that minimum metal overlap of any contact is 1λ. We want one metal₁ box to cover the column of two contacts. Then metal will be 1+2+1=4λ wide, and 1+2+3+2+1=9λ high. This satisfies Rule 7.3 (Figure 106).

•Click *Layr* and change the active layer to MET1 29.

•Click *Obj*. Double click Box to change the current object to BOX.

•Click *Add*. Add a 4×9 metal₁ box with (x,y) corners at (0.5, 2.5) and (4.5, 11.5). A blue box with a diagonal pattern appears.

•Click *Get*. Activate the metal₁ box.

•Click *Cpy*. Copy the box from corner (0.5, 2.5) to corner (8.5, 2.5).

•Click *Put*. Deactivate the metal₁ box.

•Click *Save* in the upper tool bar. Follow instructions.

This completes transistor drawing M1V14x2 (Figure 106).

Problem 102 Draw the Lasi layout for the transistor in Figure 107.

1.6 Create a cell named M1V07x2 Here is a way to create cell M1V07x2 using cell M1V14x2. (Stating from scratch using *Load* may be as easy.)

•Click *List*. Double click M1V14x2 to open it in the drawing screen.
•Click *Save* in the upper tool bar to get the Save dialog box. Type M1V07x2 in the Save edit box. Click on OK. This creates cell M1V07x2, which is a copy of M1V14x2 that is to be converted to M1V07x2.

1.7 Modify the drawing to layout a transistor whose W/L=7λ/2λ
•Click *List*. Double click M1V07x2 to open it in the drawing screen (the goal is Figure 107).
•Click *Get*. Carefully enclose only part of the top line of poly with a rubber band box and click to make ONLY the top edge white. Repeat for active. You only have to enclose part of a line to select it. For corrections use *Put* to deselect a line.
•Click *Mov* to move the selected active line so that W is reduced to 7 (the poly edge follows). Click on the line at y=14, move the cursor to y=7 and click. The line moves down from y=14 to y=7.
•Click *Put*. Enclose the white lines in a rubber band box and click.
•Click *Get*. Select the two upper contact boxes.
•Click *Del*. This deletes the two upper contact boxes.
•Click *Get*. Select the top lines of the two metal₁ boxes.
•Click *Mov*. Move the top metal lines down to y=6.5 where metal overlaps the contact boxes by one lambda.
•Click *Get*. Select the metal and contact boxes.
•Click *Mov*. Center the selected boxes in the vertical direction of the active box. Metal₁ top line at y=5.5.
•Click *Put*. Enclose the white lines in a rubber band box and click.
•Click *Save* in the upper tool bar. Follow instructions.
This completes transistor drawing M1V07x2 (Figure 107).

Figure 107 Transistor M1V07x2 with W/L=7/2

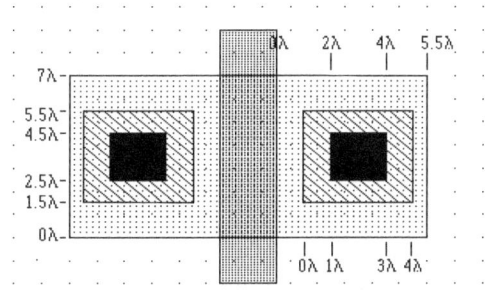

1.3.2 Place a PMOS Transistor in a Layout

The circuit layout process

2.1 Cell	2.2 Transistor	2.3 Text
2.4 N well	2.5 P select	

2.1 Create Cell Fig108

•Click *Load*, type in the cell name Fig108, Click *OK*. Rank dialog box appears. Type 2. Click *OK*. A screen with the origin marker and grid appears.

2.2 Add a pmos transistor to Fig108

Transistors are rank 1 cell drawings. They can be added to the rank 2 cell Fig108.

•Click *Obj*. Double click M1V14x2 to change the current object to M1V14x2. See lower status bar.

•Click *Add*. A box representing the transistor cell M1V14x2 appears. The transistor cell is shown in outline form with a + marking the cell's origin. Position the transistor cell origin at (0, 0) and click. The cell appears in full color.

2.3 Write Text Label the transistor.

•Click *Text*. Click on position (1,14) on the top edge of the active area of the transistor (Figure 108). Type mp1 in the TEXT dialog box, and click OK. Text appears. To set text size click *Tsiz* in *menu2*.

2.4 Draw an N well box

Industry has standardized on p wafers. This means you need an n island in the p wafer, an n well, for pmos transistors. This is a area on a p wafer that has been doped with a surplus of electrons (Section 1.1). The pmos transistor mp1 requires an n well.

Rule	Well	Lambda
1.1	Minimum width	12
1.2	Minimum spacing between wells at different potential	18
1.3	Minimum spacing between wells at samepotential	6
1.4	Minimum spacing between wells of different type	0

Well Rule 1.1 specifies minimum width of a well is 12λ. The pmos active dimensions are greater than 12λ so Rule 1.1 is satisfied.

Rule	Active	Lambda
2.1	Minimum width	3
2.2	Minimum spacing	3
2.3	Source/drain active to well edge	6
2.4	Surface/well contact active to well edge	3
2.5	Minimum spacing between active of different implant	4

How do the transistor's active and poly boxes relate to the well? The poly rules do not mention the well, however Active Rule 2.3 states that source/drain active to well edge distance is 6λ.

Figure 108 Boxes defining transistor n well and p-select

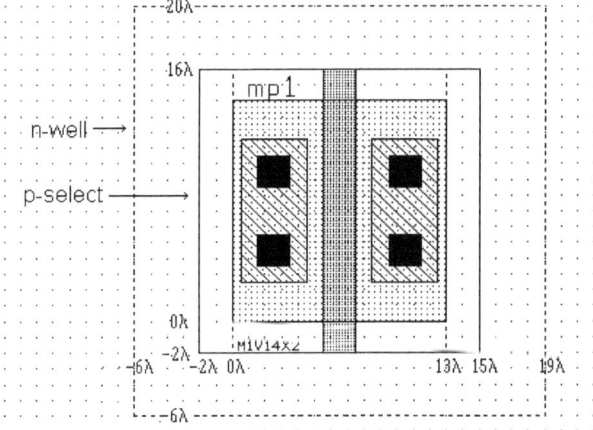

•Click *Layr* and change the active layer to NWEL 42. If necessary shrink the transistor drawing (click on *Xpnd*). (status line does not change)
•Click *Obj*. Double click Box to change the current object to BOX.
•Click *Add* (to add an n well box enclosing mp1's 14λ×13λ active area). Note that only now does the status line reflect BOX and layer NWEL. Place the box edges six lambdas from the active edges by clicking on grid dot at (−6, −6) and then at (19, 20). A dashed blue box outline appears.

2.5 Draw a P-Select box An active area is defined as p when it is enclosed by a p-select box (Figure 108).

Rule	Select	Lambda
4.1	Minimum select spacing to channel of transistor to ensure adequate source-drain width	3
4.2	Minimum select overlap of active	2
4.3	Min select overlap of contact	1
4.4	Minimum select width and spacing (P and N select may be coincident, no overlap)	2

Select Rule 4.2 specifies minimum select overlap of active is 2λ. Reminder: poly also overlaps active by 2λ (the gate extension). This helps when you draw the select box.

•Click *Layr* and change the active layer to PSEL 45.
•Click *Obj*. Double click Box to change the current object to BOX.
•Click *Add* (to add a p-select box enclosing mp1). Place the box edges two lambda's from the active edges by clicking on grid dot at $(-2, -2)$ and then at $(15, 16)$. A solid yellow box outline appears.
•Click *Save* in the upper tool bar. Follow instructions.

This completes pmos transistor drawing Fig108.

A practical matter When there are a number of pmos transistors in a circuit, one n well box and one p select box usually surround clusters of pmos transistors. The same applies to n select boxes surrounding nmos transistors.

Problem 103 Draw the Lasi layout for the transistor in Figure 108.

Problem 104 Draw the Lasi layout for the transistor in Figure 109.

Problem 105 Draw the Lasi layout for the connection in Figure 110.

1.3.3 Place an NMOS Transistor in a Layout

The circuit layout process
3.1 Cell   3.2 Transistor   3.3 Text
3.4 P well   3.5 N select

3.1 Create Cell Fig109
•Click *Load*, type in the cell name Fig109, click *OK*. Rank dialog box appears. Type 2. Click *OK*. A screen with the origin marker and grid appears.

3.2 Add an nmos transistor Rank 1 transistors can be added to the rank 2 cell Fig109. Make cell M1V07x2 the current object. Click on *Add*. A box representing the transistor cell appears. The transistor cell is shown in outlined form with a + marking the cell's origin. Position the transistor cell origin at (0, 0) and click. The cell appears in full color.

Figure 109 Transistor M1V07X2

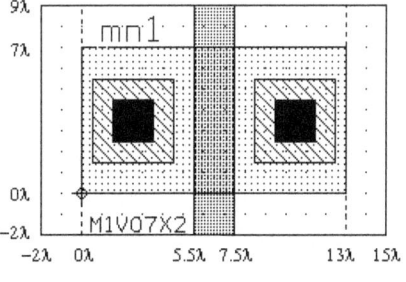

3.3 Write Text Click on *Text*. Click on the top edge of the active area. Type mn1 in the TEXT dialog box, and click on OK. Text appears.

3.4 Draw a P well box There is no need to add a p well box in a p wafer.

3.5 Add N-Select An active area is defined as n when it is enclosed with an n-select box. Select Rule 4.2 specifies minimum select overlap of active is 2λ. Reminder: poly also overlaps active by 2λ (the gate extension).

•Click *Layr* and change the active layer to NSEL 46.
•Click *Obj*. Double click Box to change the current object to BOX.
•Click *Add* (to add a n-select box enclosing mn1 whose active area is 7×13). Place the box edges two lambda's from the active edges by clicking on grid dot at (−2, −2) and then at (15, 9). A solid green outline appears.
•Click *Save* in the upper tool bar. Follow instructions.

This completes nmos transistor drawing Fig109.

1.3.4 Connect a p well in a Layout to Vss

The circuit layout process
4.1 Cell	4.2 Active	4.3 Contact
4.4 Metal1	4.5 P select	

4.1 Create Cell Fig110
•Click *Load*, type in the cell name Fig110, click *OK*. Rank dialog box appears. Type 1. Click *OK*. A screen with the origin marker and grid appears.

4.2 Draw Active Box
•Click *Layr* and select the ACTV 43 layer.
  Rule 6.1 specifies exact contact size as $2\lambda \times 2\lambda$.
  Rule 6.2 specifies minimum active overlap of contact as 1.5λ.
  Combination of rule 6.1 and rule 6.2 requires a $5\lambda \times 5\lambda$ active box.
•Click *Obj*. Double click Box to change the current object to BOX.
•Click *Add*. Add a 5×5 active box with corners at (0, 0) and (5, 5). A green box with green dot pattern appears.

4.3 Draw Contact to p surface
Connections to the surface are required. The p area must be connected to ground (V_{SS}). *Contact to any wafer area is made by a metal₁-contact-active sandwich.* (The active layer is in intimate contact with the wafer surface below the active area.)
Rule 6.1 specifies exact contact size as $2\lambda \times 2\lambda$.

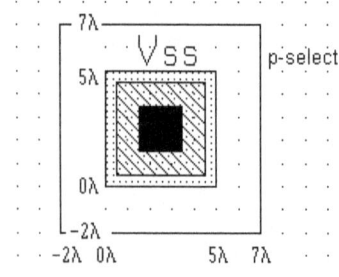

Figure 110 Connect to a p well

•Click *Layr*. Select the CONT 28 layer.
•Click *Obj*. Double click Box to change the current object to BOX.
•Click *Add*. Add a 2×2 contact box with corners at (1.5, 1.5) and (3.5, 3.5). A blue box appears.

4.4 Draw Metal₁ box
Combination of rule 7.3 and rule 6.1 requires a $4\lambda \times 4\lambda$ metal₁ box.
•Click *Layr* and select the MET1 29 layer.

•Click *Obj*. Double click Box to change the current object to BOX.
•Click *Add*. Add a 4λ×4λ metal₁ box with corners at (0.5, 0.5) and (4.5, 4.5). A blue box with a diagonal pattern appears

4.5 Draw P Select box The active box automatically connects to the p wafer. Connecting the Vss ground metal box to the p wafer via the contact and the active box requires adding a p-select box to define the active box as p+ (it's a p well).

Rule 4.2 specifies minimum select overlap of active is 2λ. The active box is 5×5. This requires a 9×9 p-select box to surround the active box with a 2λ overlap.

•Click *Layr* and select the PSEL 45 layer.
•Click *Obj*. Double click Box to change the current object to BOX.
•Click *Add*. Add a 9×9 p-select box enclosing the Vss active box with minimum 2λ select overlap of active. Corners at (−2,−2) and (7,7). A yellow box outline appears.
•Add Text V$_{SS}$ to this box.
•Click *Save* in the upper tool bar. Follow instructions.

This completes p well contact drawing Fig110.

1.3.5 Connect an n well in a Layout to V$_{dd}$

The circuit layout process (assuming you are drawing inside an n well)

5.1 Cell	5.2 Active	5.3 Contact
5.4 Metal1	5.5 N select	

5.1 Create Cell Fig111
•Click *Load*, type in the cell name Fig111, click *OK*. Rank dialog box appears. Type 1. Click *OK*. A screen with the origin marker and grid appears.

5.2 Draw Active Box
•Click on Layer and select the ACTV 43 layer.
The combination of rule 6.1 and rule 6.2 requires a 5λ×5λ active box.
•Click *Obj*. Double click Box to change the current object to BOX.
•Click *Add*. Add a 5×5 active box with corners at (0, 0) and (5, 5). A green box with green dot pattern appears.

5.3 Draw Contact to n well surface Connections to the surface are required. The n-well area, the pmos body, has to be connected to supply voltage V_{DD}.

Contact to any wafer area is made by a metal$_1$-contact-active sandwich. The active layer is in intimate contact with the wafer surface below the active area. Rule 6.1 specifies exact contact size as $2\lambda \times 2\lambda$.

•Click *Layr* and select the CONT 28 layer.
•Click *Obj*. Double click Box to change the current object to BOX.
•Click *Add*. Add a 2×2 contact box with corners at (1.5, 1.5) and (3.5, 3.5). A light blue box appears.

5.4 Draw Metal$_1$ box Rule 7.3 specifies minimum overlap of any contact as 1λ. Rule 6.1 specifies exact contact size as $2\lambda \times 2\lambda$. Combination of rule 7.3 and rule 6.1 requires a $4\lambda \times 4\lambda$ metal1 box.

•Click *Layr* and select the MET1 29 layer.
•Click *Obj*. Double click Box to change the current object to BOX.
•Click *Add*. Add a $4\lambda \times 4\lambda$ metal$_1$ box with corners at (0.5, 0.5) and (4.5, 4.5). A blue box with a diagonal pattern appears

Figure 111 Connection to a n well

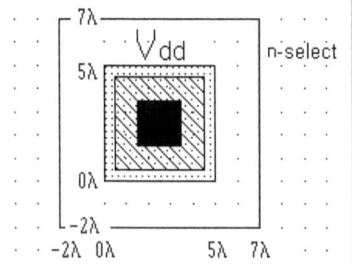

5.5 Draw N Select box The purpose of the active box is connection to the n well. Connecting the Vdd metal box to the n-well in the p wafer via the contact and the active box requires adding a n-select box to define the active box as n. Rule 4.2 specifies minimum select overlap of active is 2λ. The active box is $5\lambda \times 5\lambda$. This requires a $9\lambda \times 9\lambda$ n-select box to surround the active box with a 2λ overlap.

•Click *Layr* and select the NSEL 46 layer.
•Click *Add*. Add a $9\lambda \times 9\lambda$ n-select box enclosing the Vdd active box with minimum 2λ select overlap of active. Corners at (−2,−2) and (7,7). A green box outline appears.
•Add Text V_{DD} to this box.
•Click *Save* in the upper tool bar. Follow instructions.
This completes n well contact drawing Fig111.

1.3.6 Layout a Frame

Frame Cell One layout method uses cells of fixed height and variable width to accommodate a range of circuit sizes. Fixed height allows cells to be placed next to each other in a row so that a V_{SS} ground wire runs across the bottom of the cells, and a power supply V_{DD} wire runs across the top of the cells. An empty cell is referred to as a frame. In addition to adding power and ground wires the n well area is connected to supply voltage V_{DD}, and the p area (the p wafer) is connected to ground V_{SS}.

Allowing for the large nwell size we define a frame to have dimensions $25\lambda \times 55\lambda$. Contact to V_{SS} (1.3.4), and V_{DD} (1.3.5) is made by placing contacts and active below the V_{SS} and V_{DD} metal$_1$ wires. P-select and n-select define the active regions as p+ and n+ respectively, and an n well in the upper part of the frame defines an n region for pmos transistors.

The circuit layout process

6.1 Cell	6.2 Active	6.3 Metal$_1$	6.4 Contact
6.5 P select	6.6 N select	6.7 Outline	6.8 N Well

6.1 Create Cell Fig112
•Click *Load*, type in the cell name Fig112, click *OK*. Rank dialog box appears. Type a 2. Click *OK*. A screen with the origin marker and grid appears.

Figure 112 Frame Layout

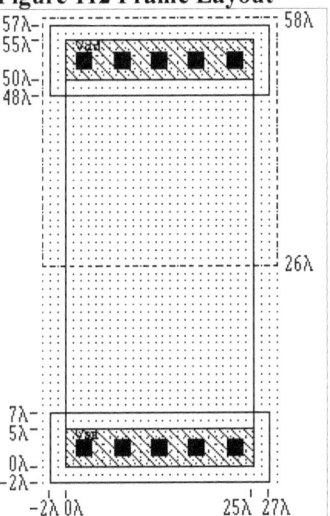

6.2 Draw Active box
Rule 6.1 specifies exact contact size as $2\lambda \times 2\lambda$. Rule 6.2 specifies minimum active overlap of contact as 1.5λ.

Combination of frame 25λ width, rule 6.1 and rule 6.2 requires a $5\lambda \times 25\lambda$ active box.

•Click *Layr* and select the ACTV 43 layer.
•Click *Obj*. Double click Box to change the current object to BOX.
•Click *Add*. Add a $25\lambda \times 5\lambda$ active box with corners at (0, 0) and (25, 5). A green box with green dot pattern appears.

•Click *Get*. Activate the active box.

•Click *Cpy*. Copy this box from corner (0, 0) to corner (0, 50). This is the active box for power supply V_{DD}. To do this click *Up* in the upper toolbar while holding *Ctrl* down. This moves the active box down screen in small steps until the upper box shows.

•Click *Put*. Deactivate the active box.

6.3 Draw Metal₁ boxes for V_{SS} and V_{DD} wires

Rule	Metal₁	Lambda
7.1	Minimum width	3
7.2	Minimum spacing	3
7.3	Min overlap of any contact	1
7.4	Minimum spacing when either metal₁ line is wider than 10λ	6

Rule 7.1 specifies minimum width as 3λ.

Rule 7.2 specifies minimum wire spacing as 3λ.

Rule 7.3 specifies minimum metal₁ overlap of any 2 × 2 contact is 1λ.

•Click *Layr* and select the MET1 29 layer. Metal₁ will connect via contacts to active. Therefore metal₁ must be 1+2+1=4λ wide or more per rule 7.3. Make it 5λ wide to cover the active box.

•Click on *Fit* and *Xpnd* to center the frame on screen.

•Click *Obj*. Double click Box to change the current object to BOX.

•Click *Add*. Add a 5λ wide×25λ long metal₁ box with corners at (0,0) and (25,5). A blue box with a diagonal pattern appears.

•Click *Add*. Add a 5λ wide×25λ long metal₁ box at the top with corners at (0,50) and (25,55). A blue box with a diagonal pattern appears.

Problem 106 Draw the Lasi layout for the connection in Figure 111.

Problem 107 Draw the Lasi layout for the frame in Figure 112.

6.4 Draw the Contact boxes

Rule	Simple Contact to Active	Lambda
6.1	Exact contact size	2x2
6.2	Minimum active overlap	1.5
6.3	Minimum contact spacing	3
6.4	Min spacing to gate of transistor	2

Rules 6.1 and 6.3 combined specify contact to contact distance from contact left edge to next contact left edge is 5λ. For 5 contacts cell width must be $5\lambda \times 5\lambda = 25\lambda$. When adjacent cells are taken into account the first and last contacts must be 1.5λ from the left and right edges. Center the 5 contact pattern in the $5\lambda \times 25\lambda$ metal$_1$ boxes.

- Click *Layr* and select the CONT 28 layer.
- Click *Obj*. Double click Box to change the current object to BOX.
- Click *Add*. Add a $2\lambda \times 2\lambda$ box with corners at (1.5, 1.5) and (3.5, 3.5). A light blue box with solid fill appears. Add one contact every 5λ.
- Click *Get*. Use Get to activate the 5 contact boxes.
- Click *Cpy*. Copy the 5 contacts from corner (0,0) to corner (0, 50).
- Click *Put*. Enclose the contact boxes with a rubber band box and click to erase white lines (deactivate).

6.5 Draw the P Select box

Rule 4.2 specifies minimum select overlap of active as 2λ.

The active box is 5×25. This requires a 9×29 p-select box surrounding the 5×25 active box with a 2λ overlap on all sides.

- Click *Layr* and select the PSEL 45 layer.
- Click *Obj*. Double click Box to change the current object to BOX.
- Click *Add*. Add a 9×29 p-select box enclosing the V_{SS} active box with minimum 2λ select overlap of active. Corners at (−2, −2) and (27,7). A yellow box outline appears.

6.6 Draw the N Select box

Rule 4.2 specifies that minimum select overlap of active is 2λ.

The active box is 5×25. This requires a 9×29 n-select box surrounding the 5×25 active box with a 2λ overlap.

•Click *Layr* and select the NSEL 46 layer.
•Click *Obj*. Double click Box to change the current object to BOX.
•Click *Add*. Add a 9×29 n-select box enclosing the V_{DD} active box with minimum 2λ select overlap of active. Corners at (−2, 48) and (27, 57). A green box outline appears.

6.7 Draw the Outline box An optional outline of the cell is useful. The corners of the frame cell are at (0,0) and (25,55). Therefore the cell outline is a 25×55 box.

•Click *Layr* and select the OUT 2 layer.
•Click *Obj*. Double click Box to change the current object to BOX.
•Click *Add*. Add a 25×55 outline box with corners at (0,0) and (25,55). An orange outline box appears.

6.8 Draw the N well box

Rule 2.4 specifies surface/well contact active to well edge as 3λ. In other words, active Rule 2.4 specifies 3λ from active edge to n well edge when the active is used for a contact from active to the well.

•Click *Layr* and select the NWEL 42 layer.
•Click *Obj*. Double click Box to change the current object to BOX.
•Click Add. Add a 31×28 n well box with corners at (−3,30) and (28,58). A dashed blue box outline appears. This makes the upper half of the frame area an n region.
•Click *Text*. Add Text V_{DD} to the upper box and Text V_{SS} to the lower box.
•Click *Save* in the upper tool bar. Follow instructions.

This completes frame drawing Fig112.

1.3.7 Layout a CMOS Inverter

The symbol for the pmos transistor mp1 (Figure 113a) has a bubble at its gate, whereas the mn1 nmos symbol does not. (We selected this pair of symbols from many possible pairs used in the literature.)

Understanding how the inverter circuit works (page 58) requires knowing how mos transistors work, and how parasitic capacitors make certain circuit behavior possible in addition to the undesirable effect of increasing the rise and fall time for rail-to-rail waveform transitions.

Figure 113a Inverter   **Figure 113b Inverter sketch**

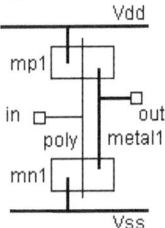

The circuit layout process

7.1. Sketch a layout	7.2. Copy a frame	7.3. Draw transistors
7.4. Add transistors	7.5. Add Select	7.6. Draw wires/pads

7.1. Sketch a layout
The sketch (Figure 113b) uses boxes, thin lines, and thick lines to represent active, poly, and metal respectively. Contacts are not shown explicitly. In and out are metal2 terminals. Let Fig113c be the top ranking cell.

7.2. Make Fig113c a copy of frame Fig112
•Click *List*. Double click Fig112.
•Click *Get*. Activate the entire frame.
•Click *Copy* (menu 2). Type *Fig113c* in the Copy edit box. Click OK
•Click *Import* (in upper toolbar). Type *Fig113c* as filename. Click OK. It's now in *List*.
•Click System in the upper tool bar. In the system dialog box click *Rerank*. Use List to select cell name Fig113c. Type 2 as rank, and click on OK.

7.3 Draw transistors

No drawing is required, because we will use existing M1V14x2 and M1V07x2.

7.4. Add transistors

Transistors are rank 1 cell drawings. They are added to the rank 2 Fig113c cell. Here is how you do it.

7.4.1 M1V07x2

•Click *Obj* and select M1V07x2 as the object. We want the transistor's lower n-select edge to be coincident with the V_{SS} p select line so that there is no gap or overlap. The transistor n select line has not been added yet so use the poly lower edge as a guide.

•Click *Add* and bring the cursor onto the drawing screen. A box representing the M1V07x2 cell appears. Position the cell's origin at (5,9) and click. The M1V07x2 cell appears in full color.

7.4.2 M1V14x2

•Click *Obj* and select M1V14x2 as the object. We want the transistor's upper p-select edge to be coincident with the V_{DD} n-select line so that there is no gap or overlap. The p-select line has not been added yet so use the poly upper edge as a guide.

•Click *Add* and bring the cursor onto the drawing screen. A box representing the M1V14x2 cell appears. Position the cell's origin at (5,32) and click. The M1V14x2 cell appears in full color.

7.4.3 Write Text
Label the transistors so that we can talk about them.
•Click *Text*. Click on a position on the top edge of the active area of the nmos transistor. Type mn1 in the TEXT dialog box, and click OK. Red text appears. Repeat for pmos transistor mp1.

7.4.4 N well
The frame has an N well (1.3.6 Frame Layout).
Active Rule 2.3 states that source/drain active to well edge distance is 6λ. Check that the lower edge of the N well is 6λ from mp1 active, and 6λ or more from mn1 active. Use *Mov* to move it as required.

7.5 Add Select

An active area is defined as p+ when it is enclosed with a p select box.
An active area is defined as n+ when it is enclosed with a n select box.

Rule 4.2 specifies that minimum select overlap of active is 2λ. Reminder: poly also "overlaps" active by 2λ. This helps when you draw the select box.

Rule	Select	Lambda
4.1	Minimum select spacing to channel of transistor to ensure adequate source-drain width	3
4.2	Minimum select overlap of active	2
4.3	Min select overlap of contact	1
4.4	Minimum select width and spacing (P and N select may be coincident, no overlap)	2

P Select

•Click on *Obj.* Change the current object and BOX.

•Click *Layr.* Select the PSEL 45 layer.

•Click *Add.* Add a box overlapping mp1's active area by two lambdas. This makes mp1 a pmos transistor. Box corners are at (3,30) and (20,48).

N Select

•Click *Layr* and change the current layer to NSEL 46.

•Click *Add.* Add a box overlapping mn1's active area by two lambdas. Mn1 is now an nmos transistor. Box corners are at (3,7) and (20,18).

7.6. Draw wires and frame i/o pads.

A poly wire connects the gates.

A metal$_1$ box connects mn$_1$ source to V_{SS}.
A metal$_1$ box connects mp$_1$ source to V_{DD}.
A metal$_1$ box connects mp$_1$ drain to mn$_1$ drain.

A poly-contact-metal$_1$ pad connected to the gates is the input terminal.
A metal$_1$-via$_1$-metal$_2$ pad connects the input terminal to the outside world.
A metal$_1$-via$_1$-metal$_2$ pad connects the output terminal to the outside world.

7.6.1 Poly

Rule	Poly	Lambda
3.1	Minimum width	2
3.2	Minimum spacing over field	3
3.2a	Minimum spacing over active	3
3.3	Min gate extension of active	2
3.4	Min active extension of poly	3
3.5	Minimum field poly to active	1

Rule 3.1 specifies minimum poly width as 2λ. Rule 3.5 specifies minimum field poly to active edge spacing as 1λ.

•Click *Layr* and change the active layer to POL1 44.
•Click *Add.* Connect gates. Add a 2×12 poly box with (x,y) corners at (10.5,18) and (12.5,30).
•The in i/o pad. *Add* a 5×5 poly box with (x,y) corners at (10.5,19) and (5.5,24).

7.6.2 Metal 1 We have not distinguished between source and drain, because the transistor active areas on either side of the poly gate stripe are identical as drawn. The choice of source or drain is made when you wire the transistor into the circuit. The choice does not matter, because the transistor layout is symmetrical.

Reference sketch (Figure 113b).

Rule	Metal$_1$	Lambda
7.1	Minimum width	3
7.2	Minimum spacing	3
7.3	Min overlap of any contact	1
7.4	Minimum spacing when either metal$_1$ line is wider than 10λ	6

Rule 7.1 specifies minimum width as 3λ. Therefore metal lines are 3λ wide when not over a contact. However 3λ metal lines that extend over a contact might as well be 4λ wide.

•Click *Layr* and change the active layer to MET1 29.
•Click *Add*.

Figure 113c Inverter Layout

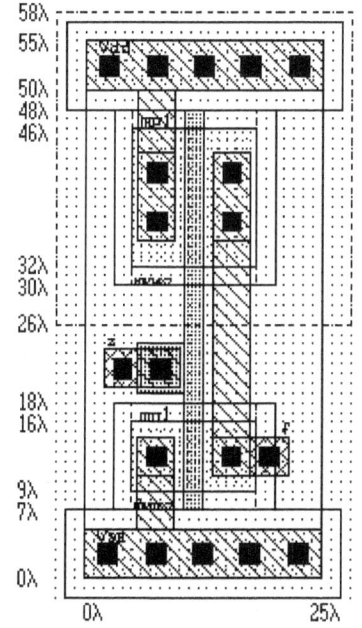

I/o pads
•Add a $4\lambda \times 4\lambda$ metal$_1$ box centered over the poly square with corners at $(10, 19.5)$ to $(6, 23.5)$.

•At the left hand side of the poly 5×5 square Add a $4\lambda \times 4\lambda$ metal$_1$ box adjoining the right hand edge of the poly's metal$_1$ with corners at $(6, 19.5)$ and $(2, 23.5)$.

•At the right hand side of mn$_1$ Add a $4\lambda \times 4\lambda$ metal1 box adjoining the right hand edge of mn$_1$'s metal$_1$ with corners at $(17.5, 10.5)$ and $(21.5, 14.5)$.

•Add Text to identify these i/o pads as input x, and output f. The logic equation is f=x′.

Wire the inverter transistors
•Add a metal$_1$ box to connect mn$_1$ source (the left side of mn$_1$) to V_{SS} with (x,y) corners at $(9.5, 10.5)$ and $(5.5, 5)$.

•Add a metal$_1$ box connecting mn$_1$ drain to mp$_1$ drain (the right sides of mn$_1$ and mp$_1$) with (x,y) corners at $(13.5, 14.5)$ and $(17.5, 34.5)$.

•Add a metal$_1$ box to connect mp$_1$ source (the left side of mp$_1$) to V_{DD} with (x,y) corners at $(9.5, 50)$ and $(5.5, 43.5)$.

Click *View*. Click on *Clear all* layers. Uncheck *All Layers*.
Check MET1 29. Click *OK*. Verify all metal$_1$ pieces are connected.
Click *View*. Click *Set All* layers. Click *OK*. View the inverter.

Problem 108 Draw the Lasi layout for the inverter in Figure 113.

7.6.3 Contact The input line is poly so you need a contact to metal1.

Rule	Simple Contact to Poly	Lambda
5.1	Exact contact size	2x2
5.2	Minimum poly overlap	1.5
5.3	Minimum contact spacing	3
5.4	Minimum spacing to gate of transistor	2

Rule 5.2 specifies a 1.5λ contact overlap by poly,

Rule 5.1 specifies a 2λ×2λ contact,

Rule 7.3 specifies a 1λ contact overlap by metal$_1$.

We have a sandwich: 5×5 poly box, 2×2 contact box, and 4×4 metal$_1$ box.

Mosis Design Rules 6. Simple Contact to Active. Rule 6.1 specifies exact contact size is 2λ×2λ.

•Click *Layr* and change the active layer to CONT 28.

•Click *Add*. Add a 2λ×2λ contact box with (x,y) corners at (7, 20.5) and (9, 22.5) centered in the input terminal's 5λ×5λ poly box.

Add a 2λ×2λ contact box with (x,y) corners at (18.5, 11.5) and (20.5, 13.5) centered in the output terminal's 4λ×4λ metal$_1$ box..

7.6.4 Via$_1$ A metal$_1$/via$_1$/metal$_2$ sandwich is needed at input z & output f.

Rule	Via 1	Lambda
8.1	Exact size	2x2
8.2	Min via1 spacing	3
8.3	Min overlap by metal1	1
8.4	Min spacing to via1 when stacked vias are not allowed	2
8.5	Min spacing to poly and active	2

•Click *Layr* and change the active layer to VIA1 30.

•Click *Add*. Add a 2×2 via$_1$ box with lower left corner at (18.5, 11.5) centered in the output i/o pad's 4×4 metal$_1$ box.

•Click *Add*. Add a 2×2 via$_1$ box with lower left corner at (3, 20.5) centered in the input i/o pad's 4×4 metal$_1$ box.

7.6.5 Metal2 We need a metal$_1$ to metal$_2$ output pad sandwich of 4×4 metal$_1$ box, 2×2 via$_1$ box, and 4×4 metal$_2$ box. And, we need a metal$_1$ to metal$_2$ sandwich for the input pad.

Rule	Metal$_2$	Lambda
9.1	Minimum width	3
9.2	Minimum spacing to metal2	3
9.3	Minimum overlap of Via1	1
9.4	Min spacing when either line is wider than 10λ	6

•Click *Layr* and change the active layer to MET2 31.
•Click *Add*.
•Add a metal$_2$ box with corner at (17.5, 10.5) centered over the output terminal's 4×4 metal$_1$ box.
•Add a metal$_2$ box with corner at (6, 19.5) centered over the input terminal's 4×4 metal$_1$ box.

This completes inverter drawing Fig113c.

1.4 Electromigration

Process specifications – Background – Other failure mechanisms –
Packaged devices – Failure rate – MTTF (Mean time to one failure)
Electron volt – Example

Process specifications The electromigration phenomena restricts current
densities in integrated circuit metal layers. Typical current density is cast
into practical dimensions closely related to circuit layout dimensions.

$$Typical\ current\ density\ :\ 1E5\ \frac{A}{cm^2} = 10^5\ \frac{A}{cm^2}\frac{1cm^2}{10^8\mu m^2} = 1\frac{mA}{\mu m^2}$$

Here are typical DC current specifications per μm width of metal layer,
and perimeter of contacts and vias per design rules.

layer	I_{max}	Stacks	I_{max}
Contact	1.0mA	contact – via1	0.6mA
Metal 1	1.0mA / μm	via1 – via2	0.4mA
Via 1	0.6mA	via2 – via3	0.6mA
Metal 2	1.0mA / μm	via1 – via2 – via3	0.6mA
Via 2	0.6mA		
Metal 3	1.0mA / μm		
Via 3	1.0mA		
Metal 4	1.5mA / μm		

Background Electromigration is the transport of metal ions through a
conductor.[1] At high current densities, there is significant momentum
transfer from electrons to metal atoms. This removes metal from the
highest current density point in the conductor which increases conductor
resistance and further increases current density. Ultimately, this process
leads to an open conductor. Electromigration is among the most widely
known and extensively researched failure mechanisms for IC's. As
temperature and current density increase metal ions migrate more readily.
Resistance to migration depends on conductor properties such as atomic
mass, melting point and grain structure. Resistance to migration also
depends on dimensions of the structure, and on the layers below and above
the conductor.

---

[1] W. Burger, C. Christenson, D. Murray, S. Yazzie 1998, "Aluminum-Based
Metallization Enhance Device Reliability", Microwave Journal (Oct): LC Call
Number: TK7800.M5

Gold has a very high atomic mass (197) compared to aluminum's atomic mass (27) or to copper's atomic mass (63.5). Gold's melting point is 1063 °C compared to aluminum's 658 °C. These properties make gold extremely resistant to electromigration. Other properties also favor gold over aluminum: resistivity is 18% lower, thermal conductivity is 36% higher, and coefficient of thermal expansion is 40% lower.

All metal layer systems require a good barrier to prevent diffusion into silicon. At 370 °C, gold and silicon form a eutectic alloy that leads to leaky and shorted junctions. Aluminum and silicon withstand a somewhat higher temperature (the eutectic alloy forms at 577 °C). At lower temperatures, however, aluminum also mixes into silicon causing problems with junctions and contacts. Adding silicon to aluminum reduces the problem, but adding a barrier layer (e.g. TiN) is preferred. Sometimes both are done.

A metal layer system must adhere to underlying materials (e.g. silicon dioxide). Aluminum atoms adhere to oxide, while gold requires an additional layer to ensure good adhesion. Most gold metal layer systems use the diffusion barrier to meet this additional requirement. For example, one gold system uses TiW(N) both as a barrier and an adhesion layer.

Reliability of metal interconnects is measured by lifetime experiments on a set of metal lines. Lines are stressed at high current densities ($10\text{mA}/\mu\text{m}^2$ or greater) and at elevated temperatures (150 °C to 250 °C). A common criteria for time to failure is open circuit for single layer structures, or 20% increase in resistance for multiple layers.

Time to failure is plotted on a log linear graph, and values of T_{50} (time for 50% of the samples to fail) is extracted together with the standard deviation. The general expression for mean time to (one) failure (MTTF), first proposed by Black (page 37), is:

$$MTTF = Ai^{-n}e^{\frac{E_a}{kT}}$$

where
A = material constant based on microstructure and geometric properties
i = conductor current density in A/cm^2
n = constant whose value ranges between 1 and 3 (2 is often used for aluminum)
E_a = activation energy (eV), e.g. 0.52 eV
k = 1.3805E–23 the Boltzmann constant
T = temperature (degrees K)

Constants A, n and E_a are determined to predict conductor lifetime at any current or temperature. This requires data at three different temperatures and three different currents (three conditions confirm acceleration results from a single failure mechanism). Most experimenters use standard test structures, so their results, when carefully reported, are useful to others. For example, based on generally accepted values from the literature, gold metallized, 2 GHz power transistors have roughly 100 times longer MTTF than similar aluminum metallized devices.

Other failure mechanisms Electromigration in wires is not the only mechanism affecting circuit lifetime. Contacts to the semiconductor migrate at high currents, so this interface must also include a barrier layer that prevents metal from mixing with the semiconductor (see prior paragraphs). A semiconductor device itself may degrade. For example, threshold voltage (V_T) drift caused by hot carrier injection is most serious for CMOS, while degradation of bipolar power devices is not common under normal operation conditions. Other material related degradation generally accelerates exponentially as a function of temperature.

It is important to assess reliability under worst case, high temperature conditions. Gold's better thermal conductivity helps spread heat, and its lower electrical resistance reduces dissipation, thus lowering junction and conductor temperatures.

Packaged devices An all gold system has significant advantages in packaged devices. Aluminum metal layers lead to potential problems associated with bond wires and the interface between aluminum and gold. First consider the interface. Purple plague ($AuAl_2$ intermetallic formation) results from long term Al-Au mixing that makes metal brittle so that it ultimately fails. The phenomena is more pronounced at elevated temperatures. Consequently aluminum and gold are inadvertently mixed. This problem is completely avoided if transistors use gold metallized chips and gold bond wires. When a die has aluminum metal layer, selecting a low temperature gold aluminum interface minimizes purple plague. For example, it is better to bond aluminum wire to a gold plated package rather than gold wire to the aluminum metal layers on a transistor chip.

Aluminum bond wires, however, lead to additional problems. Aluminum has higher electrical and thermal resistance than gold so it has higher ohmic loss and reaches a higher temperature. Further, its thermal expansion coefficient is higher than gold's. This causes problems when amplifying signals having time varying waveforms. Wires heat and cool in response to a signal's amplitude resulting in wire movement that causes fatigue. Failure occurs when a wire's ends work harden. Under some circumstances, failure occurs in just a few months. Gold's lower thermal and electrical resistance, lower expansion coefficient and better malleability combine to make it very resistant to wire fatigue.

Finally, since aluminum is chemically active, some failures occur when it reacts with oxygen or contaminants. A reaction is more likely to occur at high temperature. For example, bond wires connecting a chip's output terminals to a package or output capacitor are generally hotter and therefore more likely to react (oxidize). Oxidation reduces a wire's current handling capacity, further increasing temperature so that eventually, wires may melt. In comparison, gold tolerates contaminants, and its higher thermal conductivity results in lower temperature. Both are significant advantages.

Failure rate The failure rate unit is the *fit*.

$$1 fit = \frac{1\,failure}{10^9\,hours\ of\ operation}\,(definition)$$

MTTF Mean time to one failure MTTF under dc stress is described by Black's equation.[2][3] See prior paragraph.

Electron volt The energy represented by an electron accelerating through a potential difference of 1 volt is referred to as one electron volt.

$$definition: 1\,coulomb = 6.241847 \times 10^{18}\,electrons$$

$$definition: 1\,volt = \frac{1\,joule}{1\,coulomb}$$

$$energy = charge \times volt\ so\ that$$

$$1\,eV = 1\,electron \times 1\,volt = 1\,electron \times \frac{1\,joule}{1\,coulomb}$$

$$1\,eV = \frac{1e}{6.241847 \times 10^{18}\,e} \times 1\,joule = 0.160209 \times 10^{-18}\,joules$$

---

[2] Black, J. R. 1969. "Electromigration - A brief survey and some recent results" IEEE Transactions on Electron Devices Vol. ED-16, no. 4: p338
LC Call Number: TK7870.I2
[3] Black, J. R. 1969. "Electromigration Failure Modes in Aluminum Metallization for Semiconductor Devices" IEEE Proceedings Vol. 57, no. 4: p1587 LC Call Number: TK7700.I6

CMOS Circuit Design

Example

$$\frac{E_a}{kT} = \frac{0.52eV \times 0.160209 \times 10^{-18}\,\frac{J}{eV}}{1.3805 \times 10^{-23}\,\frac{J}{deg} \times 273\,deg} = 22.105$$

$$MTTF = Ai^{-n}e^{\frac{E_a}{kT}} = 10^9 \frac{1}{i^{1.5}} \ hours \ where \ i \ is \ in \ \frac{mA}{\mu m^2}$$

$$MTTF = 10^9 \ hours \ for \ 1 \ fit, \ I = 1\frac{mA}{\mu m^2} \ \ and \ \ n = 1.5$$

$$find\,A: \quad 10^9 = A \times 1^{-1.5} \times e^{22.105} = A \times 1 \times 3.982 \times 10^9 \quad \Rightarrow \quad A = 0.251$$

$$Check: MTTF = AI^{-1.5}e^{\frac{E_a}{kT}} = 0.251 \times 1 \times e^{22.105}$$

$$= 0.251 \times 1 \times 3.982 \times 10^9 = 1 \times 10^9 \ hours$$

1.5 Mosis rules

Rule	Well	Lambda
1.1	Minimum width	12
1.2	Minimum spacing between wells at different potential	18
1.3	Minimum spacing between wells at samepotential	6
1.4	Minimum spacing between wells of different type	0

Rule	Active	Lambda
2.1	Minimum width	3
2.2	Minimum spacing	3
2.3	Source/drain active to well edge	6
2.4	Surface/well contact active to well edge	3
2.5	Minimum spacing between active of different implant	4

Rule	Poly	Lambda
3.1	Minimum width	2
3.2	Minimum spacing over field	3
3.2a	Minimum spacing over active	3
3.3	Min gate extension of active	2
3.4	Min active extension of poly	3
3.5	Minimum field poly to active	1

Rule	Select	Lambda
4.1	Minimum select spacing to channel of transistor to ensure adequate source-drain width	3
4.2	Minimum select overlap of active	2
4.3	Min select overlap of contact	1
4.4	Minimum select width and spacing (P and N select may be coincident, no overlap)	2

Rule	Simple Contact to Poly	Lambda
5.1	Exact contact size	2x2
5.2	Minimum poly overlap	1.5
5.3	Minimum contact spacing	3
5.4	Minimum spacing to gate of transistor	2

Rule	Simple Contact to Active	Lambda
6.1	Exact contact size	2x2
6.2	Minimum active overlap	1.5
6.3	Minimum contact spacing	3
6.4	Min spacing to gate of transistor	2

Rule	Metal$_1$	Lambda
7.1	Minimum width	3
7.2	Minimum spacing	3
7.3	Min overlap of any contact	1
7.4	Minimum spacing when either metal$_1$ line is wider than 10λ	6

Rule	Via 1	Lambda
8.1	Exact size	2x2
8.2	Min via1 spacing	3
8.3	Min overlap by metal1	1
8.4	Min spacing to via1 when stacked vias are not allowed	2
8.5	Min spacing to poly and active	2

Rule	Metal$_2$	Lambda
9.1	Minimum width	3
9.2	Minimum spacing to metal2	3
9.3	Minimum overlap of Via1	1
9.4	Min spacing when either line is wider than 10λ	6

2 MOS nmos and pmos Transistors

The metal oxide semiconductor (MOS) transistor has two implementations, nmos and pmos, of the drain d, gate g, and source s regions (Figure 200 which is not to scale). We explain the *how*, but not the *why* of MOS transistors. We refer you to semi-conductor texts that explain the why of MOS device physics. (In modern MOS transistors the metal gate is replaced by a polysilicon gate.)

Figure 200 nmos & pmos transistors

The goal of a design is selection of values for channel length L, width W, and the correct value of mobility μ for each transistor.

The MOS transistor is a *voltage to current* amplifying device. The *transconductance* g_m is expressed as $i_{DS}=g_m v_{GS}$. A graphical display of $g_m v_{GS}$ is a I_{DS},V_{DS} plot of transistor drain current (I_{DS}) as a function of drain/source voltage (V_{DS}) with gate to source voltage (V_{GS}) as a parameter. This is the MOS *vi constraint* (Figure 20211, page 43). The concept of active device *loadline* shows us how to bias the MOS transistor for linear operations. This is why a loadline is superimposed on the MOS *vi* constraint (Figure 20111, page 43).

The MOS transistor has three operating regions:
(1) the MOS is *turned off* when $V_{GS}<V_T$ the threshold voltage,
(2) the triode region is defined by $0<V_{DS}<V_{DSsat}$ $V_{DSsat} = (V_{GS}-V_T)$,
(3) the (linear) saturation region is defined by $V_{DS}>V_{DSsat}$.

The MOS device is non-linear. Spice is designed to analyze MOS as well as BJT non-linear devices. The MOS Spice models used here are the University of California Predictive Technology Model Beta Version. The PTM model cards and BSIM models are available at

Figure 203 MOS layout

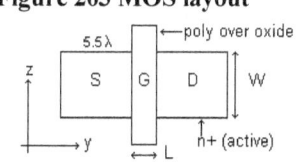

http://ptm.asu.edu/
http://www-device.eecs.berkeley.edu/bsim/

2.1 MOS Transistor vi constraint

Any component has a vi constraint. The resistor vi constraint is defined by Ohm's law v=iR. There is no parameter influencing the resistor's vi constraint. Active MOS devices are different, because they have one parameter, gate voltage, influencing the vi constraint. MOS devices produce a change in drain current when the gate voltage changes.

The output current flows into a load impedance that is a resistor R. The R defines a loadline (Figure 20111 page 43). The loadline idea applies to all active device circuit designs. Once the loadline is superimposed on an active device's vi constraint the DC operating point can be established so that dynamic inputs (sine waves, pulses, etc) are amplified without distortion by the transistor.

MOS transistor vi constraint   An MOS gate-source voltage V_{GS} is the parameter in the MOS vi constraint. There is one I_{DS},V_{DS} plot for each value of constant gate voltage V_{GS}. The I_{DS},V_{DS} plots for V_{GS} equal to 0.8V and 1.8V show the control V_{GS} exerts (Figure 20111).

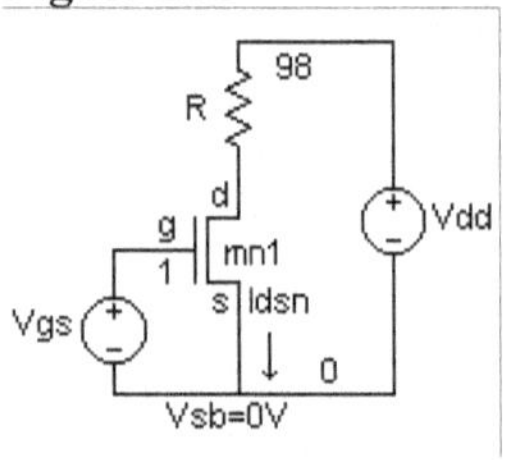

Points on a loadline Resistor R connected from a 1.8volt V_{DD} supply to the drain is referred to as a load resistor. The resistor's loadline is plotted by calculating points A and D. If $V_{GS}=0$ the transistor is off, and $V_{DS}=1.8V$ (point A Figure 20111). If the transistor is shorted out of the circuit, then $I_{DS}=1.8V/1.2K=1.5mA$ (point D). If $V_{GS}=0.8V$ the drain current is 0.3ma, the iR drop is 1.2K×0.3mA or 0.36V so that V_{DS} is about 1.44V (point B). If $V_{GS} = 1.8V$, then *this time we take the point C coordinates from Figure 20111* to get a graphical solution. The drain current is about 1mA, the iR drop is 1.2K×1mA or 1.2V and so V_{DS} is about 0.6V (point C).

Figure 201

MOS transistors in a Spice program A transistor has 4 terminals drain d, gate g, source s, and body b. It is entered into a spice program as follows. The fake numbers 4,3,2,1 are terminal pin numbers.

name	d	g	s	b	model	L	W	
MN1	4	3	2	1	N1	L=0.18u	W=1.8u	; 20/2=W/L

or as a subckt (page 48).

Spice program 2011

```
Fig2021.ckt  mos vi constraints
Vdd  98  0  DC 0
V1    1  0  DC 0
.include 180_N1P1.txt
MN1 98 1  0  0 N1  L=0.18u  W=1.8u  ; 20/2=W/L
.DC LIN VDD 0 1.8 0.05 LIN V1 -0.2 1.8 1
.PLOT DC ID(MN1) 0,1.5M
.end
```

Figure 20111 Nmos Transistor Collector Current vi Constraint

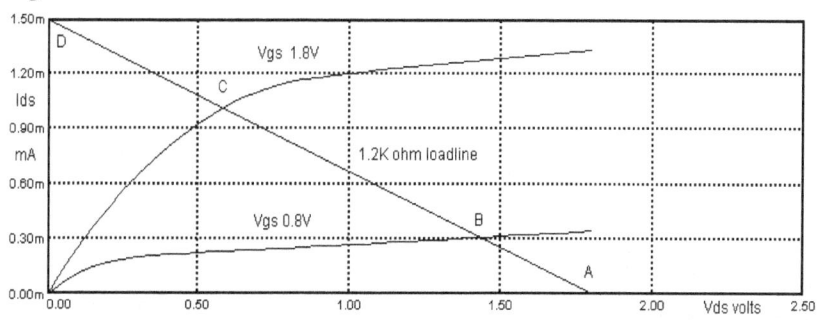

Spice program 2021

```
Fig2021.ckt  mos vi characteristics
V1  1 0 DC 0
V2  2 0 DC 0
.include 180_N1P1.txt
MN1 2 1 0  0 N1  L=0.18u  W=1.8u  ; 20/2=W/L
.DC LIN V2 0 1.8 0.05 LIN V1 0.4 1.8 0.2
.TEMP 27
.PLOT DC ID(MN1) 0,1.5M
.end
```

Figure 20211 Nmos Transistor Collector Current vi Constraint

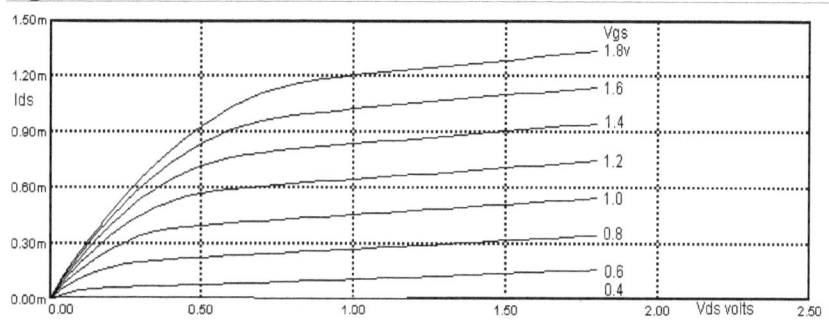

2.2 MOS Transistor DC Equations

Assume the *nmos* source is connected to ground, a positive voltage V_{GS} is connected to the gate, a positive voltage V_{DS} is connected to the drain, and $V_{SB}=0$ (Figure 103). If the channel is on the x axis, then let the source be at x=0 and the drain be at x=L. The channel is W wide (z axis into the page). The channel current I_{DS} has the same value at any place in the channel (all electrons leaving the source arrive at the drain).

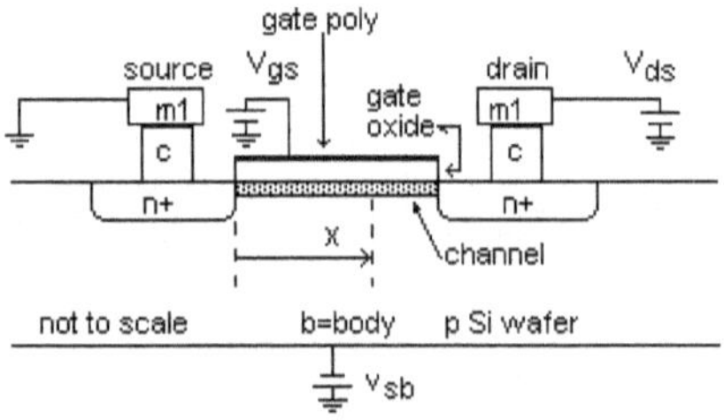

The gate area is defined by the W×L area of a conducting sheet of polysilicon (poly) material over the silicon dioxide layer of thickness t_{ox}. Voltage V_{GS} is connected from gate to source. The voltage on the gate V_{GS} draws electrons with negative charge into the channel. V_{DS} creates the

Figure 103 NMOS cross section

electric field in the channel from drain to source that imparts a velocity to the electrons causing current to flow.

Long Channel Equations

region	equation	condition
cutoff	$(1a)\ I_{DS}=0$	$V_{GS}<V_T$
triode	$(1b)\ I_{DS}=\beta[(V_{GS}-V_T)V_{DS}-0.5V_{DS}^2]$	$0<V_{DS}\leq(V_{GS}-V_T)$
saturation	$(1c)\ I_{DS}=\dfrac{\beta}{2}(V_{GS}-V_T)^2$	$(V_{GS}-V_T)\leq V_{DS}$
beta	$(1d)\ \beta=\mu_n C_{ox}\dfrac{W}{L}=\mu_n\dfrac{\varepsilon_{ox}}{t_{ox}}\dfrac{W}{L}$	

Short Channel Equations

region	equation	condition
cutoff	$(2a)\ I_{DS}=0$	$V_{GS}<V_T$
triode	$(2b)\ I_{DS}=\dfrac{1}{1+\dfrac{V_{DS}}{E_cL_e}}\mu_{eff}\dfrac{W}{L_e}C_{ox}[(V_{GS}-V_T)-\frac{1}{2}V_{DS}]V_{DS}$ *or a form that eliminates* μ_{eff} $(2c)\ I_{DS}=\dfrac{2}{E_cL_e+V_{DS}}v_{sat}WC_{ox}[(V_{GS}-V_T)-\frac{1}{2}V_{DS}]V_{DS}$	$V_{DS}\leq V_{DSsat}$
sat	$(2d)\ I_{DS}=v_{sat}WC_{ox}[V_{GS}-V_T-V_{DSsat}]$	$V_{DS}\geq V_{DSsat}$

2.3 Drain Current I_{DS} vs W

DC analysis Spice program 2045 (text file Fig2045.ckt) defines a series of nmos transistors with increasing width W, while holding channel length L constant. At $V_{DD}=0.8V$ we get

$$(3a)\ i_{drive\_n} = \frac{I_{dsn}}{W} = \frac{3.3mA}{5.4\mu m} = 0.61\frac{mA}{\mu m} \quad (3b)\ i_{drive\_p} = \frac{I_{dsp}}{W} = \frac{1.4mA}{5.4\mu m} = 0.26\frac{mA}{\mu m}$$

Figure 20451 Nmos vi constraints for various W (ΔW=1.35µm), L=0.18µm

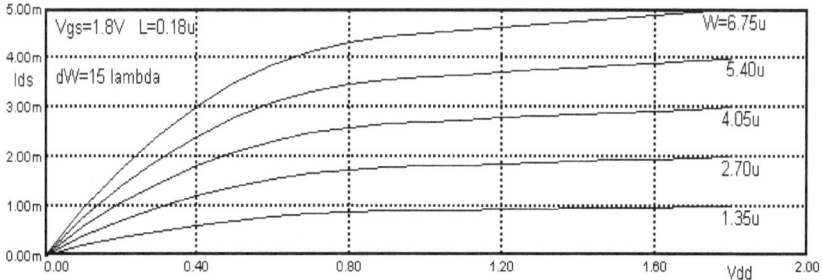

Figure 20471 Pmos vi constraints for various W (ΔW=1.35µm), L=0.18µm

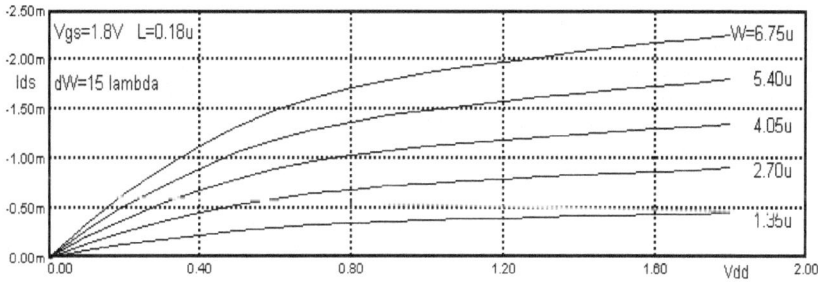

Spice program 2045

```
Fig2045.ckt nmos increasing W
MN3 98 20 0 0 N1 L=0.18u W= 1.35u      ; W=15λ
MN4 98 20 0 0 N1 L=0.18u W= 2.70u      ; W=30λ
MN5 98 20 0 0 N1 L=0.18u W= 4.05u      ; W=45λ
MN6 98 20 0 0 N1 L=0.18u W= 5.40u      ; W=60λ
MN7 98 20 0 0 N1 L=0.18u W= 6.75u      ; W=75λ
Vgs 20 0  DC 1.8
Vdd 98 0  DC 0
.include 180_N1P1.txt
.DC LIN VDD 0 1.8 0.1
.TEMP 27
.PLOT DC ID(MN3) ID(MN4) ID(MN5) ID(MN6) ID(MN7) 0,5M
.end
```

2.4 Adding Drain and Source Capacitance to the MOS Model

In Chapter 1 we learned that the design rules specify source and drain areas to be $W\lambda \times 5.5\lambda = 5.5W\lambda^2$ (Figure 203).

Figure 202 MOS capacitors

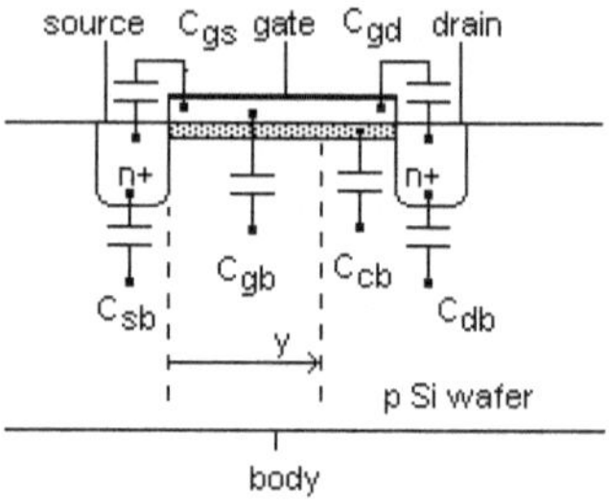

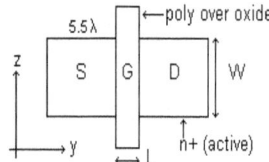

Here we show how to calculate the parameters AD, AS (channel area drain and source), & PD, PS (channel perimeter drain and source) for the MOS models used by Spice. The parameters add the drain to body (substrate) C_{db} and source to body C_{sb} to the Spice model.

Figure 203 MOS Layout

AD, AS areas of drain or source diffusion channel bottoms is $W\lambda \times 5.5\lambda = 5.5W \lambda^2$.

PD, PS perimeters of drain or source channel sidewalls is $2(W\lambda+5.5\lambda) = 2(5.5+W)\lambda$.

In Spice notation, picometers p equals micrometers μ squared. The short channel process lambda equals 0.09μm.

Example: $L = 0.18\mu m = 2\lambda$, $W_p = 1.08\mu m = 12\lambda$, $W_n = 0.45\mu m = 5\lambda$

If $W_n = 5\lambda$, then

$AD_n = AS_n = 5.5W\lambda^2 = 5.5 \times 5 \times 0.09^2 = 0.223\mu m^2$

$PD_n = PS_n = 2(5.5+W)\lambda = 2(5.5+5) \times 0.09 = 1.89\mu m$

If $W_p = 12\lambda$, then

$AD_p = AS_p = 5.5W\lambda^2 = 5.5 \times 12 \times 0.09^2 = 0.535\mu m^2$

$PD_p = PS_p = 2(5.5+W)\lambda = 2(5.5+12) \times 0.09 = 3.15\mu m$

For example:
```
MN1  103 102 101 99 N1  L=0.18u  W=0.45u          ; 5λ/2λ=W/L
+ AD=0.223p   AS=0.223p    PD=1.89u   PS=1.89u

MP3  209 208 207 298 P1  L=0.18u  W=1.08u          ; 12λ/2λ=W/L
+ AD=0.535p   AS=0.535p    PD=3.15u   PS=3.15u
```

The Spice Subckt, MOS Models, MOS Circuits and the MOS Body

In Spice programs we simplify program writing by introducing two levels of abstraction using *subckts* for transistors, and *subckts* for circuits using those transistors.

Our transistor sub circuit (subckt) definitions are stored in files mos018_2.lib (L=2λ=0.18μm), and mos018_L.lib (L=10λ=0.90μm). Appendix 2 page 186. Create subckts if you need other W values.

Any MOS integrated circuit design uses some set of MOS transistors with various widths W in one or more circuits. Writing Spice programs is simplified if each transistor is defined as a sub circuit. The contents of the mos018_2.lib file are included in any Spice program by the dot lib (.lib) control statement when you run the Spice program. For example

```
        .lib mos018_2.lib
or      .lib C:\ElectricCircuits\cm1spice\mos018_2.lib
```

Subcircuit definitions start with *.subckt*, and end with *.ends <sub circuit name>* or just *.ends* depending on your version of Spice. All statements between *.subckt* and *.ends* are included in the definition (page 48).

The control statement syntax that calls a sub circuit in a Spice program must begin with an x or X such as

x2 n1 n2 n3 ... nn <sub circuit name>

where the number of nodes n1 n2 n3 ... nn is the same as the number of nodes in the sub circuit definition.

A note about the body terminal The purpose of the body connection is to reverse bias, turn off, the wafer to channel parasitic diode.

In a digital circuit suppose the power supply voltages are +1.8V and 0V. Then a pmos body is connected to +1.8v, and nmos body to 0V.

In an analog circuit suppose the power supply voltages are +1.8V, 0V, and −1.8V. If the circuit is a differential amplifier with gates at 0V, then a pmos body is connected to +1.8v, and nmos body to −1.8V. Repeat −1.8V.

Calling a subckt in a Spice program The control statement syntax that uses sub circuits in a Spice program must start with an x or X. For example:

```
.include 180_N1P1.txt   ; transistor models N1, P1
.lib mos018_2.lib       ; transistor subckts mp1, mn13
*name   nodes              subckt      size
XMP1      15   5   7   7   mp1    ;   W/L=5/2, node 7 is +V     page 188
XMN1      15   5   0   0   mn13   ;   W/L=100/2, node 0 is 0V.  page 187
```

The number of nodes in XMP1 and XMN1 is the same as the number of nodes in the sub circuit definitions in mos018_2.lib. The transistor *.subckts* use models N1 and P1 that are defined in 180_N1P1.txt (Section 2.5). The node *numbers* in the program may be different from the node *numbers* in the dot subckt definition. The W/L comments in lambda (λ) units are for the user's convenience.

Example First level of abstraction Transistors
$$\frac{W}{L} = \frac{4.5\mu m}{0.18\mu m} = \frac{50\times 0.09}{2\times 0.09} = \frac{50\lambda}{2\lambda} = 25 \quad where \quad \lambda = 0.09\mu m$$

```
*subckt name drain gate source body nodes
.subckt mn8 124  123  122   99      ;nmos transistor
      MN8  124  123  122  99 N1 L=0.18u W=4.50u    ;50/2=W/L
      + AD=2.2275p  AS=2.2275p  PD=9.99u  PS=9.99u
.ends mn8

.subckt mp8 224  223  222  298      ;pmos transistor
      MP8  224  223  222  298 P1 L=0.18u W=4.50u ;50/2=W/L
      + AD=2.2275p  AS=2.2275p  PD=9.99u  PS=9.99u
.ends mp8
```

Figure 204

Example Second level of abstraction
```
.subckt fig204 2 6 5 7 9
xmp1 4 4 7 7 mp8    ; no x on mp8
xmp2 5 4 7 7 mp8    ; 7 is V7=1.8V
xmn1 4 2 3 9 mn8    ; body at node 9
xmn2 5 6 3 9 mn8    ; 9 is V9=-1.8V
.ends fig204
```

The second level example is a subckt for a differential amplifier. We take names of subckts (fig204) from their circuit figures. The listed nodes are inputs, outputs, and power supplies.

2.5 The MOS Spice Models

MOS transistor models are silicon wafer fabrication process dependent. For each process there are models for nmos and pmos transistors. Two major variables, channel length L and channel width W, size any transistor. The BSIM models for the L=0.18μm pmos and nmos devices are in Appendix 2 page 184. There are two major applications of MOS transistors. One is integrated circuits. The other is the huge infrastructure of discrete *high power* MOS devices, which we do not discuss.

Lambda A process can resolve some minimum length unit that is referred to as lambda λ (This is the L_{min} λ, not the r_{out} λ). All dimensions in the *design rules* are some multiple of λ so that layouts are not made obsolete by reductions in λ. By design the

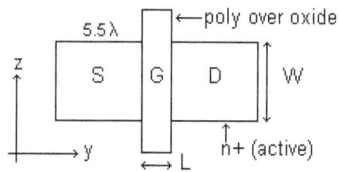

Figure 203 Transistor Layout

minimum channel length L is $L_{min}=2\lambda$. Transistors laid out with L_{min} are short channel devices in these modern processes. *A fact to keep in mind is that when longer L such as $5L_{min}$ are used in a layout the transistors have the properties of a long channel device.* The other dimension defining transistor size is channel width W. The greater the channel width W the greater the current the transistor can deliver. The switching speed of an MOS transistor is proportional to its channel length $L=n\lambda$ ($L_{min}=2\lambda$). As L decreases the time required for the transistor to switch on and off decreases. Consequently the chip industry has created a sequence of silicon wafer fabrication processes based upon shorter and shorter channel lengths L_{min}. In turn, new Spice models for nmos and pmos transistors were developed for each process. Note: The BSIM model cards show $L_{max}=L_{min}$. Spice programs correctly calculate performance for any practical selection of $L>L_{min}$.

Layout Laying out a circuit intended to be fabricated by a process requires adhering to a set of *design rules* for that process (e.g. www.mosis.com). The rules specify minimum allowable values for various lengths, widths, spacing, overlaps, and so forth. When these rules are correctly implemented the processing during wafer fabrication is *independent* of the layout patterns.

Mobility Short Channel

Velocity model In long channel devices the mobility μ is equal to the ratio of electron or hole velocity v to electric field E.

Mobility μ is the *slope* of the velocity/electric field plot (Figure M301).

Velocity is proportional to electric field when $E/E_c<1$. Then μ is constant and MOS devices are long. When E/E_c exceeds 1 electron & hole drift velocity saturates, mobility decreases to zero, and MOS devices are short. E.g. if $V_{DS}=1.8V$, $L=0.18\mu m$, then $E=1.8/0.18 =10V/\mu m$ so that $E/E_c=1.2$ ($E_c=8.33$ V/μm).

Figure M301 Velocity v in an electric field v/1.38×10^{11} μm per sec

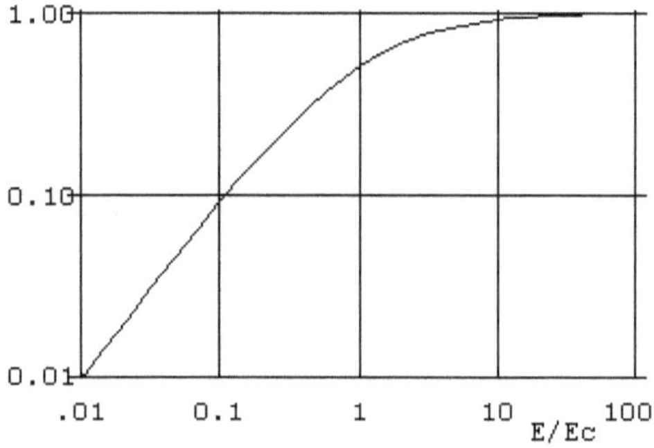

Definition of Saturation Drain Voltage V_{DSsat} is the drain voltage at which the electron (hole) current saturates in long or short transistors.

$$V_{DSsat \atop short} = \frac{(V_{GS}-V_T)E_c L_e}{E_c L_e +(V_{GS}-V_T)}$$

$$V_{DSsat \, long} = V_{GS}-V_T$$

Effective Mobility The value of μ_{eff} is estimated as follows. First, the effective transverse electric field at the surface is calculated.

$E_{t,eff} = \frac{1}{\varepsilon_{Si}}\left(\frac{Q_{inv}}{2}+Q_B\right)$ where Q_B is the bulk charge and the inversion channel charge is $Q_{inv}= C_{ox}(V_{GS}-V_T)$. The threshold voltage is

$V_T = -V_a+\frac{Q_B}{C_{ox}}$ $\Rightarrow$ $Q_B = C_{ox}(V_a+V_T)$ where $V_a=0.5V$ for typical n+

polysilicon gate devices so that $E_{t,eff} = \frac{C_{ox}}{\varepsilon_{Si}}\left(\frac{V_{GS}-V_T}{2}+V_a+V_T\right)$ and

$E_{t,eff} = \frac{V_{GS}-V_T}{6t_{ox}}+\frac{V_a+V_T}{3t_{ox}}$ where $\frac{C_{ox}}{\varepsilon_{Si}} = \frac{\varepsilon_{ox}}{\varepsilon_{Si}t_{ox}} = \frac{1}{3t_{ox}}$

Then the effective mobility is calculated, and plotted (Figure M302 and Figure M303). Empirical constants μ_1, E_1, and n are used to facilitate formation of analytical expressions

	$\mu_1 \dfrac{cm^2}{Vs}$	$E_1 \dfrac{10^6 V}{cm}$	n
electron	670	0.67	1.6
hole	290	0.35	1.0

$$\frac{1}{\mu_{eff}} = \frac{1}{\mu_1} + \frac{1}{\mu_2} \quad \Rightarrow \quad \mu_{eff} = \frac{\mu_1 \mu_2}{\mu_1 + \mu_2} = \mu_1 \frac{1}{1 + \dfrac{\mu_1}{\mu_2}} = \mu_1 \frac{1}{1 + \dfrac{E_{t,eff}^n}{E_1^n}}$$

where μ_1, E_1, and n are empirical constants

Reference
K. Toh, P. Ko, and R. Meyer 1988. An Engineering model for Short-Channel MOS Devices. *IEEE Journal of Solid-State Circuits* Vol. 23, No. 4 (August): 950-958, LC Call Number: TK7871.85 .I23

Figure M302 Effective electron mobility cm^2/Vs

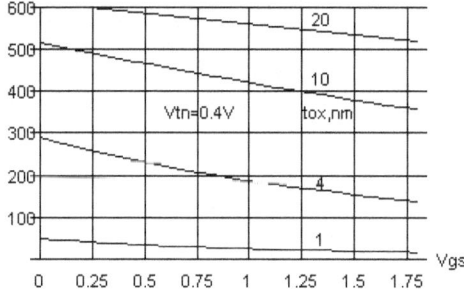

Figure M303 Effective hole mobility cm^2/Vs

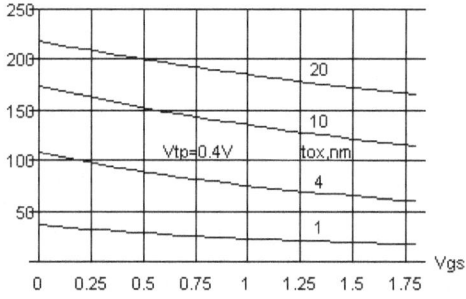

2.6 MOS Transistor Layouts

A transistor layout has four components: active box, gate poly box, contact box, and metal₁ box.

Figure 203 Basic Transistor layout

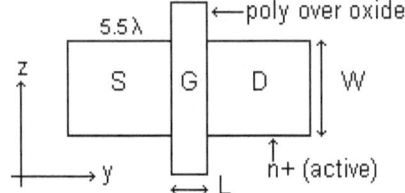

We are using numbers from pages 39, 40.

The 5.5 number (1.5+2+2) arises from rules 6.1, 6.2, 6.4 page 40.
The 6 number (2+2+2) arises from rules 6.1, 6.4 page 40.
The 3 number (3) arises from rule 3.2a page 39.

The active box height is Wλ. The active box width depends on the number of gates, and whether transistors are in series or parallel.

One gate - width is 5.5+L+5.5 λ.

Two gates in series - width is 5.5+ L+3+L+5.5 λ. (add 3+L for a gate)

Two gates in parallel - width is 5.5+ L+2+2+2+L+5.5 λ. (add 6+L)

Figure One_Gate1 (dots 1λ)

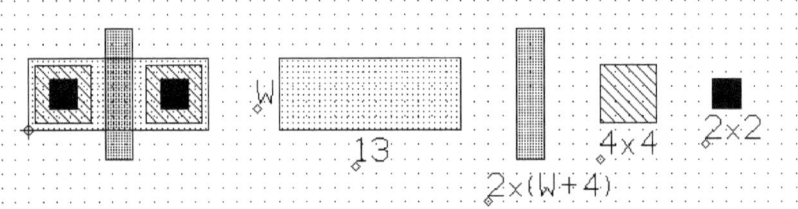

The gate poly box dimensions are L×(W+4). The 4 number is from rule 3.3.
The contact box dimensions are 2×2 per rule 6.1 page 40.
The metal₁ box over contact dimensions are 4×4 per rule 7.3 page 40.

Figure Two_Gate1 (dots 1λ)

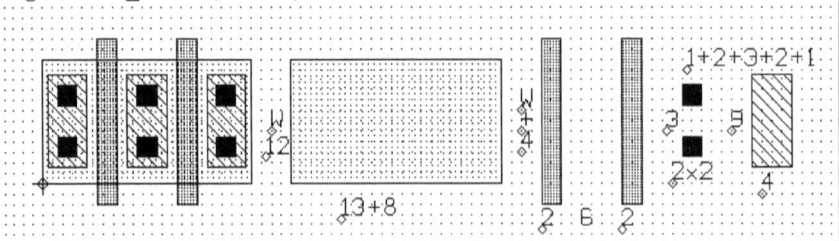

Figure Three_Gate1 (dots 1λ)

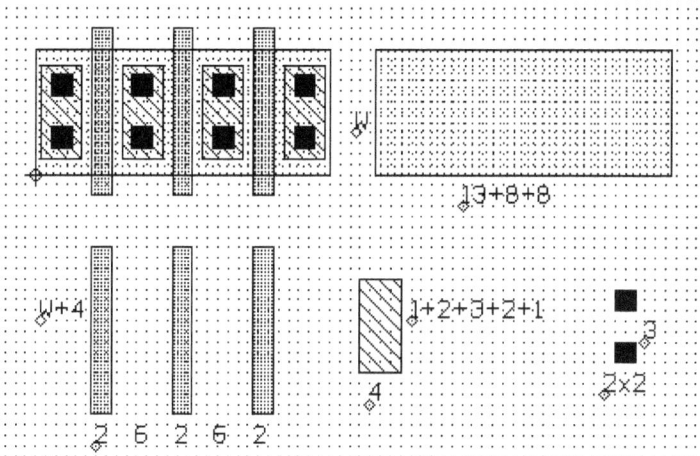

Figure One_Gate2 (dots 1λ)

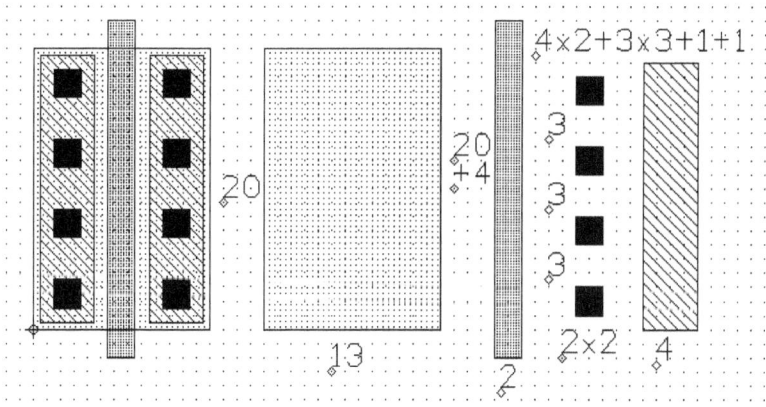

Figure One_Gate3 (dots 1λ)

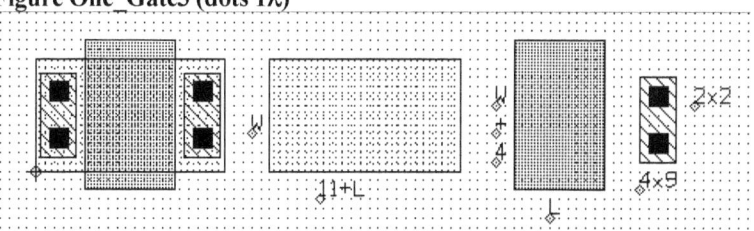

Problem 201 Layout the 5 transistors shown here.

2.7 MOS Connections and Spacings

The elements of connection boxes can be stacked one on top of the other.

The contact size is 2×2 (rule 5.1), which determines poly is 5×5 (rule 5.2), and metal$_1$ is 4×4 (rule 7.3). Rules are on pages 39 and 40.

Figure Connect1 (dots 1λ)

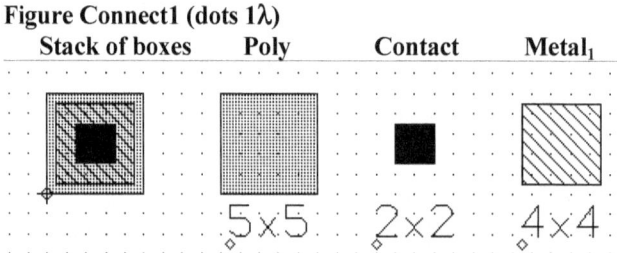

The contact size is 2×2 (rule 5.1), poly is 5×5 (rule 5.2), metal$_1$ is 4×4 (rule 7.3), via$_1$ is 2×2 (rule 8.1), metal2 is 4×4 (rule 9.3).

Figure Connect2 (dots 1λ)

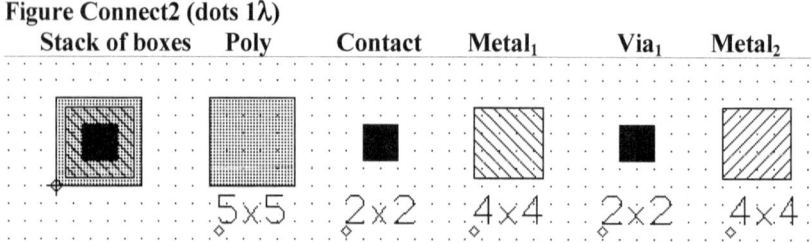

The via$_1$ size is 2×2 (rule 8.1), metal$_1$ is 4×4 (rule 7.3), metal2 is 4×4 (rule 9.3).

Figure Connect2 (dots 1λ)

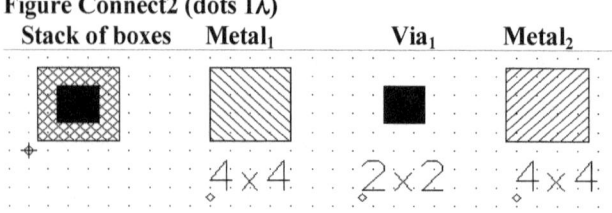

Poly to poly spacing is 3λ (rule 3.2). Poly to active spacing is 1λ (rule 3.5). The poly size is 5×5 (rule 5.2), active is any size. Spacings are as shown.

Figure Spacing1 (dots 1λ)

poly   poly   poly   active

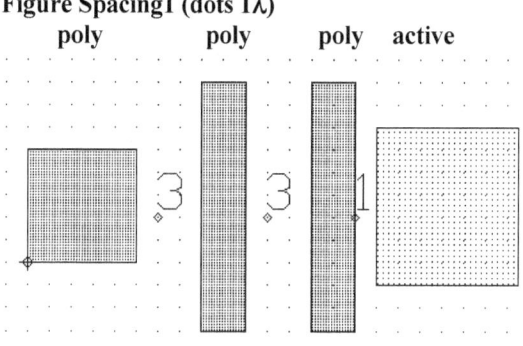

Metal to metal spacing is 3λ. Observe metal is 0.5 λ from poly box edge.

Figure Spacing2 (dots 1λ)

poly/metal₁   metal₁   metal₁

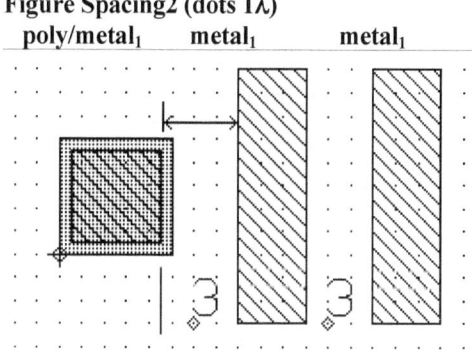

Problem 202 Layout the 5 connections and spacings shown here.

---

3 Digital Circuits with no memory

Circuits with no memory implement logic equations as OR of ANDs that is a sum of *minterms* such as $m_1+m_3+m_7$ or AND of ORs that is a product of *maxterms* such as $M_1M_3M_7$. A *no-memory* building block's values of output variables are computed from *present* values of input variables. A change of any input variable value produces new output variable values after a *time delay* required to compute the new values. The numerical value of this *propagation time delay* depends upon the implementing technology.

How do we calculate the W/L of the transistors? – Selection of L is straightforward. $L=L_{MIN}=2\lambda$ for minimum *digital* circuit propagation delay and switching time. Or, $L=4\lambda$, 6λ, 8λ, 10λ when higher output impedance is required. Selection of W depends on capacitance at internal and external nodes. We evaluate performance, and adjust W's accordingly.

Circuit Loads Any CMOS circuit output is connected via metal and polysilicon lines to input gates of driven circuits. These physical connections and the gates add up to a capacitor loading a circuit output. We call this capacitor C_{LOAD}.

Static CMOS Circuits There are *static* and *transmission gate (TG)* CMOS circuit classes. Static circuits have a *negative true logic* nmos branch that implements the circuit's *defining equation,* and a pmos branch that implements the *complement* of the circuit's defining equation. *For negative true logic active low f is the nmos circuit and active high logic f' is the pmos circuit. Exchange f and f' for positive true logic.*

Transmission gate circuits A CMOS transmission gate TG requires complementary signals to turn it on and off. This is why TG circuits include inverters. Some TG circuits use fewer transistors than their static counterparts.

The circuit layout process

1. Sketch a layout	2. Copy a frame	3. Draw transistors
4. Add transistors & Nwell	5. Add Selec	6. Draw wires/pads

Frame Template for circuit cells

Cell height is 75λ. Cell width expands as the layout expands. The Vdd metal$_1$ and Vss metal$_1$ wires across top and bottom of the frames connect power and ground to each cell. Metal$_1$ and poly are used for wiring inside the cell. Laying out the cell wiring is an art form as metal to metal spacing and poly to poly spacing 3λ rules are implemented. Poly wire resistance is about 8 ohms/square, whereas metal resistance is about 0.07 ohms/square. This is why vertical metal$_2$ wires are used to for external cell connections to cell inputs and outputs in various rows. Odd number metal wires are horizontal, and even number metal wires are vertical.

Figure 300 Frame Template

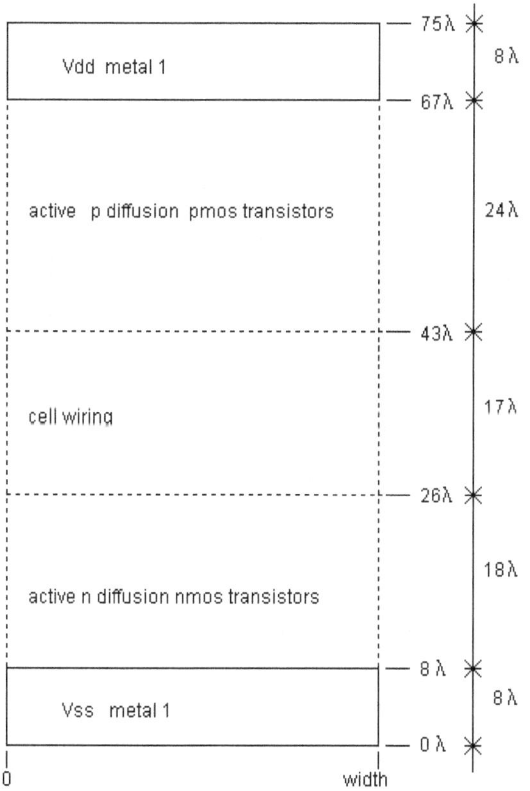

Do this after every layout is completed.
Click *View*. Click on *Clear all* layers. Uncheck *All Layers*.
Check MET1 29. Click *OK*. Verify all metal$_1$ pieces are connected.
Click *View*. Click *Set All* layers. Click *OK*. View the inverter.

3.1 NOT *Static*

The NOT circuit is the basic CMOS inverter circuit (Figure 301) where one nmos device is paired with one pmos device. When the nmos is on it discharges the output node to 0V, and when the pmos is on it charges the output node to 1.8V in a 1.8V system. Voltage transitions from 0 to 1.8V or 1.8V to 0V are referred to as rail to rail transitions. With rail to rail transitions either nmos is on or pmos is on. This is what is needed for 2-level H and L digital logic. *CMOS requires each nmos to be paired with a pmos (the standard NOT circuit is 04).*

Figure 301 CMOS inverter   **Figure 302 Inverter driving C$_{load}$**

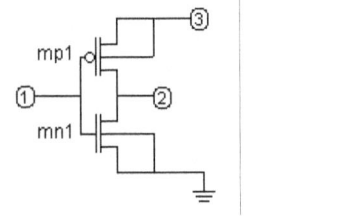   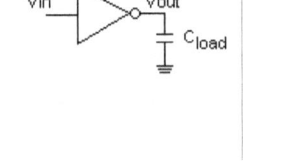

In a CMOS digital circuit C$_{LOAD}$ (Figure 302) draws a *current* i$_{LOAD}$ *only* while the applied voltage switches from L to H or H to L. For example:

(1) $i_{load} = C_{load} \dfrac{dv}{dt}$ $\Rightarrow$ $i_{load} = 100\,fF\,\dfrac{1.8V}{200\,ps} = 0.9mA$

Any MOS semiconductor process defines nmos and pmos drive current capability. The drive current process parameters for the short channel L$_{MIN}$=2λ=0.18μm process are (2.3 page 45)

(2a) $i_{ndrive} = 0.61\dfrac{mA}{\mu m}$ (2b) $i_{pdrive} = 0.26\dfrac{mA}{\mu m}$ (2c) $i_{ds} = i_{drive}\dfrac{mA}{\mu m} \times W\ \mu m$

The drain current I$_{DS}$ that is proportional to W, must be greater than or equal to i$_{LOAD}$.

(3) $W = \dfrac{i_{ds}}{i_{drive}} = \dfrac{1}{i_{drive}} i_{load} = \dfrac{1}{i_{drive}} C_{load} \dfrac{dv}{dt}$

In a 1.8V system, when C$_{LOAD}$ = 100fF, and switching time is 200pS, the inverter pmos and nmos transistors are sized as follows.

$$(4a) \quad \frac{W_n}{L} = \frac{1}{L} \frac{C_{load}}{i_{ndrive}} \frac{dv}{dt} = \frac{1}{0.18\mu m} \cdot \frac{100fF}{0.61mA/\mu m} \cdot \frac{1.8}{200ps} = \frac{1.50\mu m}{0.18\mu m} = 8.2 \rightarrow \frac{16.7\lambda}{2\lambda}$$

$$(4b) \quad \frac{W_p}{L} = \frac{i_{ndrive}}{i_{pdrive}} \frac{W_n}{L} = \frac{0.60mA}{0.26mA} \cdot \frac{17\lambda}{2\lambda} = \frac{39.2\lambda}{2\lambda} \rightarrow \frac{40\lambda}{2\lambda}$$

The ideal V_4 waveform shows that output voltage V_2 transient response waveform confirms these calculations (Figure 30311).

Figure 30111 CMOS inverter with Cload 100fF

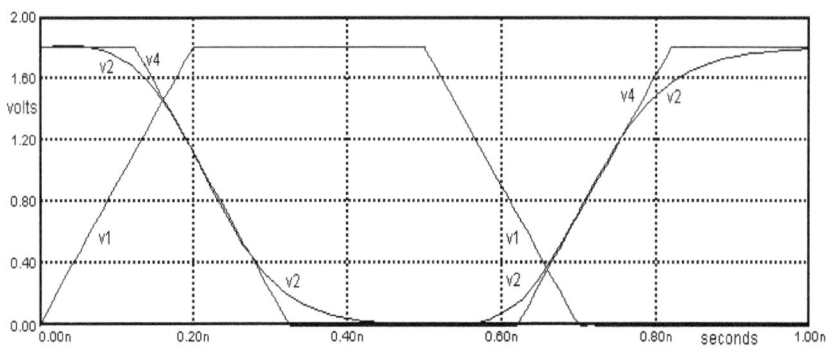

Spice program 3011

```
Fig3011.ckt  cmos inverter
Vdd 3 0 DC 1.8

V1 1 0 PULSE(0   1.8 000p 200p 200p 300p 1000p)
V4 4 0 PULSE(1.8   0 125p 200p 200p 285p 1000p)

.include 180_N1P1.txt

MP1  2 1 3 3 P1 L=0.18u  W=3.6u      ;W/L=3.6/0.18=40/2
MN1  2 1 0 0 N1 L=0.18u  W=1.5u      ;W/L=1.5/0.18=16.7/2
C2 2 0 100f

.TRAN 1e-011 1e-009 0 1e-011
.TEMP 27
.PLOT TRAN V(1) V(2) V(4) 0,2
.end
```

Problem 301 Reference Figure 301g. Draw the Lasi layouts for transistors M1V16.5x2, M3V13x2, and for the NOT where W_p=40λ & W_n=16.5λ.

NOT (inverter) Layout In order to make an integrated circuit layout we start with *layout software*, and a set of *design rules* for the process technology. Then we select the circuit we want to layout. We start with the Static Inverter NOT circuit (Figure 301).

Figure 301b **Figure 301c** **Figure 301d**

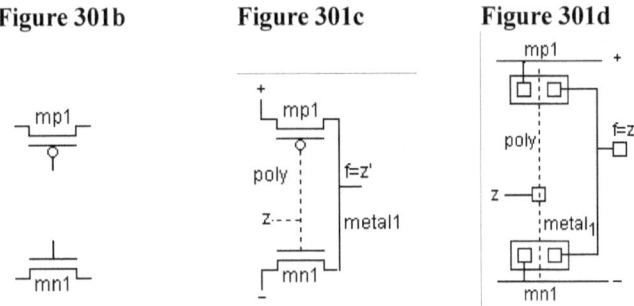

A layout plan starts with a transistor count. The inverter circuit has 1 pmos and 1 nmos transistor. The nmos and pmos transistors are positioned in rows (Figure 301b). Add poly and metal$_1$ lines defining the pmos-nmos pair as an inverter (Figure 301c). The transistor symbols are replaced by active boxes with vias and poly gates. And, input and output boxes are added (Figure 301d).

Run LASI. (LASI commands and selections are in italics.)
The available height for nmos or pmos in the frame template (page 57) is 18λ and 24λ. In this inverter $W_N=16.7\lambda$ and $W_P=40\lambda$. $W_P=40\lambda$ has to be reduced to $W_P=13\lambda$ to allow for the 24λ limit and pmos nwell box 6λ requirements (rule 1.3 page 39). This means parallel transistors are required to increase the pmos W from $W_P=13\lambda$ to $W_P=39\lambda$. Know that parallel transistor currents add.

Draw transistor M3V13x2 (Figure 301f). For three transistor poly boxes rules 3.1, 5.1, 6.1, 6.4, pages 39, 40 require active width to be 5.5+2+6+2+6+2+5.5=29λ. Add 8λ for each poly box (see pg 12 for 5.5λ).

Click on *Load*. Enter cell name *M3V13x2*. Click OK. Click *Obj* and select *box*. Click *Layer* and select *Actv 43*. At origin (0,0) add 29λ wide, 13λ high active box. Click *Layer* and select *Pol1 44*. Add three 2λ wide, 17λ high poly boxes at corners (5.5, −2), (13.5, −2), and (21.5, −2). Add 2 contacts with metal$_1$ in each active area (1.4 page 13, 1.5 page 14).

Create a transistor by adding active, poly, contact, and metal$_1$ boxes.

Draw transistor M1V16_5x2 (Figure 301e).

Click on *Load*. Enter cell name *M1V16_5x2*. Click OK. Click *Obj* and select *box*. Click *Layer* and select *Actv 43*. At origin (0,0) add 13λ wide, 16.5λ high active box. Click *Layer* and select *Pol1 44*. Add one 2λ wide, 20.5λ high poly box at corner (5.5, −2). Add 3 contacts and metal₁ in each active area (see pages 13, 14).

Figure 301e M1V16_5x2 (dots 1λ)   **Figure 301f M3V13x2 (dots 1λ)**

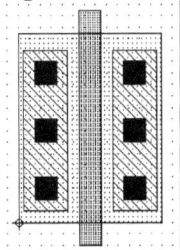   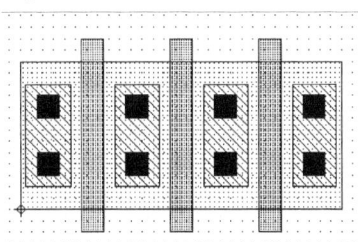

Create cell *Frame301*. Click *List*. Double click *Fig112*. *Fig112* appears on screen. *Save* as new cell *Frame301*.

Figure 301g Frame301 (dots 2λ)

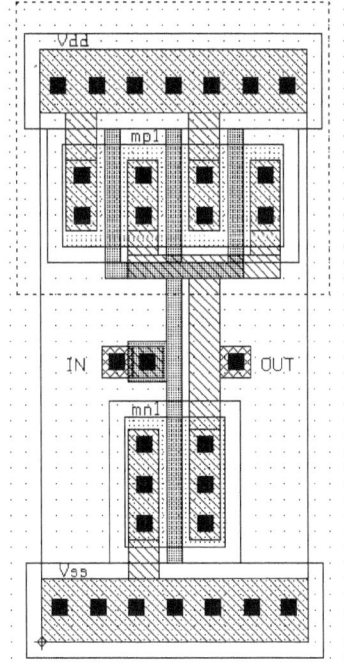

Add transistors to *Frame301* as shown in Figure 301g. To maintain 3λ active to active minimum spacing (rule 2.2 page 39) place mp1 transistor's upper p-select edge to be coincident with the V_{DD} n-select line. The p-select line has not been added yet so use the poly upper edge as a guide. Repeat for mn1 at Vss.

Clearly the frame is too narrow. *Get* the right edges in *Frame301*. *Mov* right to x=35 where outline is 3λ from transistor edge. *Put* activated lines. Add contacts to Vdd and Vss. Add p and n select (pg29).

Add poly and metal₁ wires (page 30). Add input and output boxes to the metal₂ level by using contact/metal₁ over poly and via₁/metal₂ over metal₁ (page 31).

3.2 NAND *Static*

A NAND has one output and 2 to n inputs. In practice n=2 or n=3, because transient response deteriorates as n increases. The *voltage* truth table shows the relationship of the inputs y, z and output f_{nand}. If inputs y and z are active high (H=1) and output f is active low (L=1), then f_{nand}=yz (AND). If output f_{nand} is active high (H=1), then f_{nand}=(yz)'=y'+z'. Active high and low *assignments define logic function*, while the physical behavior represented by the HL truth table does not change. (In actual practice the circuit the gate is embedded in determines the assignment[1].)

y	z	f_{nand}
L	L	H
L	H	H
H	L	H
H	H	L

How the Static Nand circuit works The circuit (Figure 303) merges two inverters so that the nmos transistors are in series, and the pmos transistors are in parallel. We perceive pairs mn_1/mp_1 and mn_2/mp_2 as inverters that have been merged. Consequently the output is L when *all* nmos are on (AND). The output is H when at least one pmos is on (OR) when the associated nmos is off (*the standard NAND circuit is 00*).

Figure 303 Static **NAND**

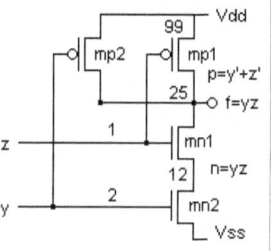

Design 2-input Static Nand

(5a) $C_{gate} = C_{ox}WL = C_{ox}\dfrac{W}{L}L^2 = \left(C_{ox}L^2\right)\dfrac{W}{L}$

(5b) $C_{gate} = 8.63\dfrac{fF}{\mu m^2}(0.18\mu m)^2\dfrac{W}{L_{min}} = 0.280\dfrac{W}{2\lambda}fF$ ($L_{min} = 2\lambda = 0.18\mu m$)

If $fanout = 5$, then $C_{load\_fF} = 5C_{gate} = 5\times 0.280\dfrac{W_{gate}}{2\lambda} = 1.4\dfrac{W_{gate}}{2\lambda}$

(6) $W_{gate} = \dfrac{2\lambda}{1.4}C_{load\_fF}$

Assume C_{load}=84fF. The nand circuit (Figure 303) shows that W_{gate} is the sum of a pmos W_p and an nmos W_n since inputs y and z are connected to nmos and pmos inputs.

(7) $W_{gate} = \dfrac{2\lambda}{1.4}84\,fF = 120\lambda = W_{p\_nand} + W_{n\_nand}$

---

[1]Nicholas L. Pappas, *Digital Design – Logic, Memory, Computers*, Section 4.2

If $W_{gate}=120\lambda$, then $C_{gate}=16.8fF$ (equation 8a), and fanout is $(84/16.8)=5$.

We also estimate C_{gate} from the gate current that flows (8b, Figure 30312).

(8a) $\quad C_{gate} = 0.280\dfrac{W}{2\lambda}\,fF = 0.280\dfrac{120\lambda}{2\lambda} = 16.8\,fF$

(8b) $\quad C_{gate} = i\dfrac{dt}{dv} = 0.35mA\dfrac{100\,ps}{1.8V} = 19.4\,fF \quad$ *(Figure 30312)*

The inverter W_n/W_p ratio is set by the ratio of the drive currents (2.3 page 45, ratio=2.35), and W is proportional to drive current. *To simplify the numbers we select 2 as the nmos and pmos drive current ratio. Then $W_p=2W_n$.* Furthermore, in a nand two nmos are in series that means each $W_{n\_nand}=2W_n$ (Problem 302). Therefore $W_{p\_nand}=W_{n\_nand}$. Let $W_{p\_nand}=60\lambda$, and $W_{n\_nand}=60\lambda$. Two basic inverters whose $W_p/W_n=60\lambda/30\lambda$ are merged.

(9a) $\quad W_p = 2W_n \qquad$ (9b) $\quad W_{n\_nand} = 2W_n = W_{p\_nand}$

(9c) $\quad W_{p\_nand} = W_{n\_nand} = \dfrac{120\lambda}{2} = 60\lambda$

Static NAND transient response Spice program 3031 transistor W's are set to 60λ. The program produces the transient response of a NAND gate (Figure 30311) with $C_{load}=5C_{in}$. Ideal waveform V_4 matches actual output V_{25} that has a propagation delay of 70ps.

Figure 30311 Two-input Static NAND, Cload/Cin=84fF/16.8fF

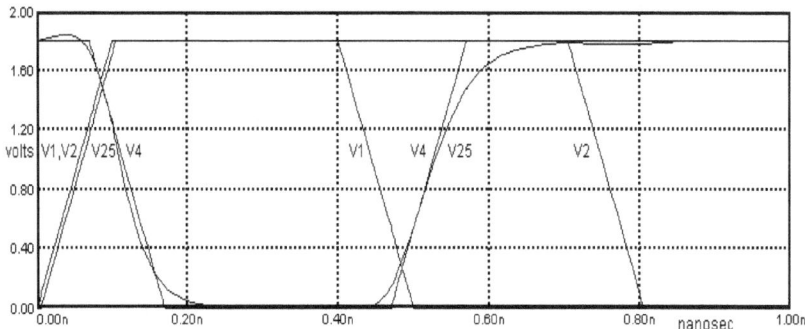

Figure 30312 Two-input Static NAND, gate currents

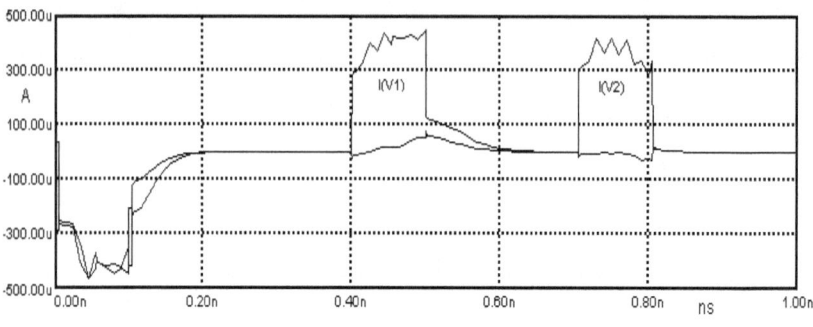

Spice program 3031 Static NAND
Fig3031.ckt   2-in static nand Transient Response

```
.include 180_N1P1.txt
Vdd 98    0 DC 1.8
Vm1 98 99 DC 0

V1 1 0 PULSE(0    1.8   000p 100p 100p 300p 1000p)
V4 4 0 PULSE(1.8   0    70p 100p 100p 300p 1000p)
V2 2 0 PULSE(0    1.8   5p 100p 100p 600p 1000p)

MP1 25  1 99  99  P1  L=0.18u  W=5.4u      ;60/2=W/L
MP2 25  2 99  99  P1  L=0.18u  W=5.4u
MN1 25  1 12   0  N1  L=0.18u  W=5.4u
MN2 12  2  0   0  N1  L=0.18u  W=5.4u
C25   25  0 84f      IC=1.8                ;Cin = 16.8f

*.PLOT TRAN V(1) V(2) V(4) V(25) 0,2
.TRAN 1e-011 1e-009 0 1e-011
.TEMP 27
.PLOT TRAN I(V1) I(V2) -500U,500U
.end
```

Problem 302 Sketch layouts of two mos in series and two mos in parallel. Deduce that parallel W's add, series L's add, and reciprocals of series W's add (think about series and parallel resistors, Figure 105 page 9).

Problem 303 Draw the Lasi layout for transistor M2V10x2. Draw the layout for a static NAND where $W_p=20\lambda$ & $W_n=20\lambda$.

NAND Layout The NAND gate circuit has 2 pmos and 2 nmos transistors. The pmos transistors are placed in a row whose area is intended to be defined as one n-well. The nmos transistors are placed in a row below the pmos transistors. (Figure 303b). The idea is to redraw the circuit schematic into a form suitable for layout.

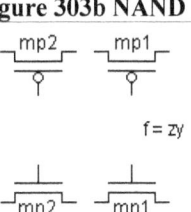

Figure 303b NAND

$f = zy$

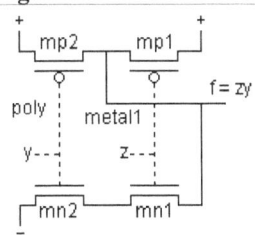

Figure 303c NAND

Add metal₁ lines to connect the outer mp₁ and mp₂ source contacts to V_{DD}. Add a metal₁ line from the drains to the output. Now mp₁ and mp₂ are in parallel (d to d to +, s to s to f out).

Add metal₁ lines to connect V_{SS} to the mn₂ source, mn₂ drain to mn₁ source, and mn₁ drain to the output. Now mn₂ and mn₁ are in series.

Add poly lines from mn₂ gates to mp₂ gates, and mn₁ gates to mp₁ gates (Figure 303c).

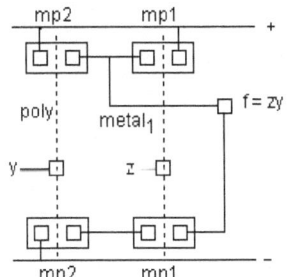

Figure 303d NAND

Poly to metal₁ connections are made by vias that are plated holes in the oxide separating the poly and metal₁ layers. Add y, z, and f vias and boxes. We use small boxes to designate vias connecting poly to metal₁ at the y and z inputs (Figure 303d).

The transistor symbols are replaced by active boxes with contacts and a poly gate.

The NAND layout needs four W=60λ transistors (2 pmos, 2 nmos). However the height of the nwell area is 24λ. So layout four parallel transistors whose W=15λ. Wiring converts the four 15λ parallel transistors into one 60λ transistor.

Layout transistor M4V15x2
Active width for one transistor is 5.5+2+5.5=13λ. For 4 transistors active width is 5.5+2+6+2+6+2+6+2+5.5=37λ. Each parallel transistor adds 2λ +6λ=8λ to active width. See page 12 for the 5.5λ number.

Click *List*. Double click *Frame300*. Save as *Frame303*. The frame is too narrow. Get frame's edges on the right side. *Mov* the right edges to the right to x=83 so that transistors will fit in the frame. Add 2 pmos and 2 nmos *M4V15x2* (Figure 303e). Observe 3λ spacing from outline to transistor active, from V_{SS}, and V_{DD} active to transistor active. Click *Text*. Add labels mp_1, mp_2, mn_1, mn_2.

Add metal1 and poly lines per Figure 303c.

Add vias and boxes per Figure 303d.

Complete the NAND layout in *Frame303*

Three vertical metal$_2$ lines connect to the outside world.

Figure 303 NAND

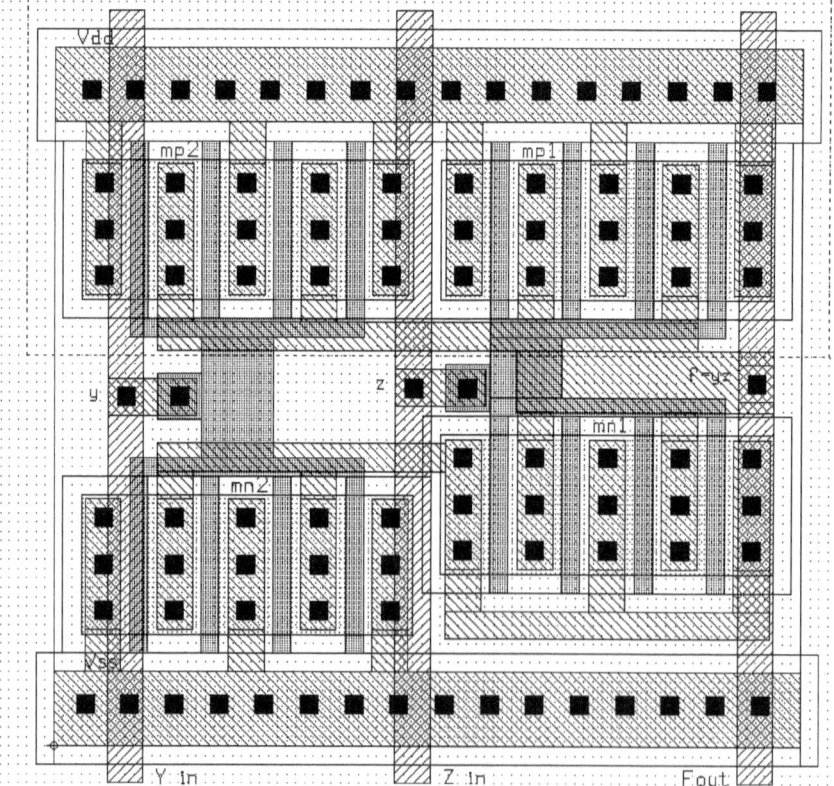

Figure 303e Frame303 (dots 1λ)

3.3 NOR *Static*

The *nor* has one output and 2 to n inputs. The *voltage* truth table shows the relationship of the inputs y, z and output f_{nor}. If inputs y and z are active high (H=1) and output f is active low (L=1), then f_{nor}= y+z (OR). If output f_{nor} is active high (H=1), then f_{nor}=(y+z)′ =y′z′ (NOR). Active high and low *assignments change logic function*, while the physical behavior represented by the HL truth table does not change. The circuit the gate is embedded in determines the assignment (*the standard NOR circuit is 02*)

Physical

y	z	f_{nor}
L	L	H
L	H	L
H	L	L
H	H	L

How the Static NOR circuit works The circuit (Figure 304) merges two inverters so that the nmos transistors are in parallel, and the pmos transistors are in series. Pairs mn_1/mp_1 and mn_2/mp_2 are merged inverters. Consequently the output is L when *either* nmos is on, and the associated pmos is off. The output is H when both pmos are on, and both nmos are off.

Figure 304 *Static* NOR

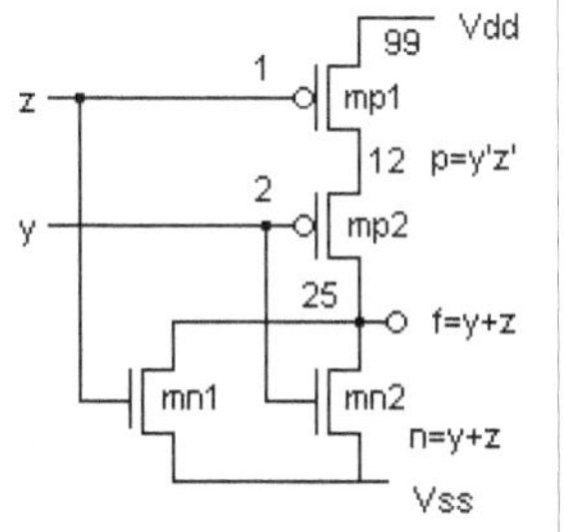

2-input Static NOR design

$$If \; fanout = 5, \; then \; C_{load\_fF} = 5C_{gate} = 5 \times 0.280 \frac{W_{gate}}{2\lambda} = 1.4 \frac{W_{gate}}{2\lambda}$$

$$(6) \; W_{gate} - \frac{2\lambda}{1.4} C_{load\_fF}$$

The W_n/W_p ratio is set by the drive current ratio (2.3 page 45), and W is proportional to drive current so that W_p=2.35W_n. However the two pmos are in series that means each W_{p_nor}=2W_p.

As a frame of reference evaluate performance of a *minimum area* NOR with smallest nmos W_n=5λ (1.5λ+2λ+1.5λ for one contact, page 40).

$$(10a) \; W_p = 2.35W_n \quad (10b) \; W_{p\_nor} = 2W_p = 2 \times 2.35W_{n\_nor} = 4.7W_{n\_nor}$$

$$(10c) \; if \; W_{n\_nor} = 5\lambda, \; then \; W_{p\_nor} = 4.7W_{n\_nor} = 4.7 \cdot 5\lambda = 23.5\lambda \approx 24\lambda$$

$$(11) \; C_{gate} = 0.280 \frac{W}{2\lambda} fF = 0.280 \frac{W_{p\_nor} + W_{n\_nor}}{2\lambda} = 0.280 \frac{29}{2} = 4.06 fF$$

CMOS Circuit Design

Assume $C_{load}=5C_{in}=5\times4.06=20fF$

$(12a)\quad i_{5\lambda} = 0.61\frac{mA}{\mu m}\times\frac{0.09\mu m}{\lambda}\times5\lambda = 0.275mA$

$(12b)\quad i = C\frac{dv}{dt}\quad\rightarrow\quad dt = \frac{C_{LOAD}}{i_{5\lambda}}dv = \frac{20\,fF}{0.275mA}1.8volts = 0.120ns$

Compare to approximate 0.1ns fall time in Figure 30421.

Static NOR transient response Spice program 3042 transistor W's are set to 24λ and 5λ. The program produces the transient response of a NOR gate (Figure 30421) with $C_{load}=5C_{in}=20fF$.

Figure 30421 *Static* NOR, Cload/Cin=20fF/4fF

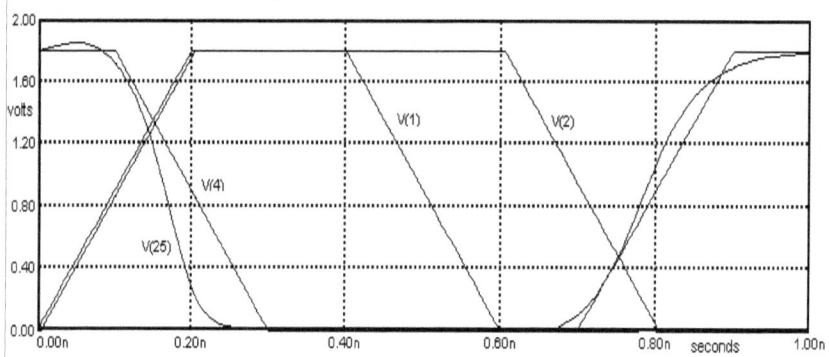

Figure 30412 *Static* NOR, gate currents

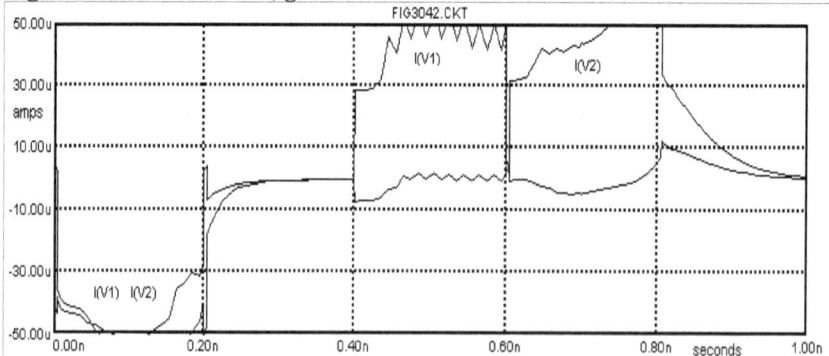

Spice Program 3042 Static NOR

Fig3042.ckt 2-in static nor Transient Response

.include 180_N1P1.txt

Vdd 99 0 DC 1.8

V1 1 0 PULSE(0   1.8   000p 200p 200p 200p 1000p)
V2 2 0 PULSE(0   1.8   5p 200p 200p 400p 1000p)
V4 4 0 PULSE(1.8   0   100p 200p 200p 400p 1000p)

MP1 99 1 12 99 P1 L=0.18u W=2.16u   ;24/2=W/L
MP2 12 2 25 99 P1 L=0.18u W=2.16u

MN1 25 1 0   0   N1 L=0.18u W=0.45u   ;5/2=W/L
MN2 25 2 0   0   N1 L=0.18u W=0.45u

C25 25 0 20f   IC=1.8   ; cin =0.28(24+5)/2=4.06f

*.PLOT TRAN V(1) V(2) V(4) V(25) 0,2
.TRAN 1e-011 1e-009 0 1e-011
.TEMP 27
.PLOT TRAN I(V1) I(V2) -50U,50U
.PRINT TRAN V(4)
.end

Problem 304 Reference Figures 304, 304g. Draw a Lasi layout for transistors M2V5x2, M2V24x2, and the two-input Static NOR where $W_p=24\lambda$ and $W_n=5\lambda$.

NOR Layout The NOR gate circuit has 2 pmos and 2 nmos transistors. The pmos transistors are placed in a row whose area is intended to be defined as one n-well. The nmos transistors are placed in a row below the pmos transistors (Figure 304b).

Figure 304b NOR **Figure 304c NOR** **Figure 304d NOR**

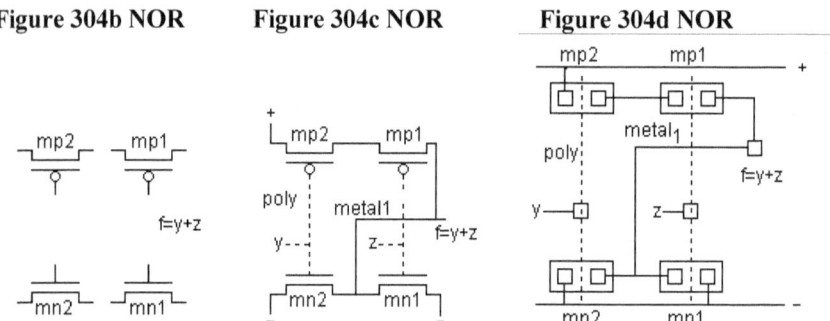

Add metal$_1$ lines to connect mp$_2$ source contact to V_{DD}. Connect the mp$_2$ drain to mp$_1$ drain (Figure 304c). Later they will be laid out in series (Figure 304f).

Add metal$_1$ lines to connect V_{SS} to mn$_2$ and mn$_1$ sources. Connect mn$_2$ and mn$_1$ drains so that they are in parallel. Connect the mp$_1$ drain to mn$_1$ drain and to the f box (Figure 304c).

Add poly lines from mn$_2$ gate to mp$_2$ gate, and mn$_1$ gate to mp$_1$ gate (Figure 304c).

Poly to metal$_1$ connections are made by vias that are plated holes in the oxide separating the poly and metal$_1$ layers. We use small boxes to designate vias connecting poly to metal$_1$ at the y and z inputs.

The transistor symbols are replaced by active boxes with contacts and a poly gate (Figure 304d).

The NOR layout needs two pmos W=24λ transistors, and two nmos W=5λ transistors.

Layout transistor *M2V24x2*.
Active width for one transistor is 5.5+2+5.5=13λ. For 2 series transistors active width is 5.5+2+3+2+5.5=18λ.

Click on *Load*. Enter cell name *M2V24x2*. Click *OK*. Click *Obj* and select *box*. Click *Layer* and select *Actv 43*. At origin (0,0) add 18λ wide, 24λ high active box. Click *Layer* and select *Pol1 44*. Add two 2λ wide, 28λ high poly boxes at corners (5.5, −2) and (10.5,−2). Select *cont 28* layer, and add 4 contacts and metal₁ in each active area. Next draw *M2V5x2*.

Figure 304 NOR

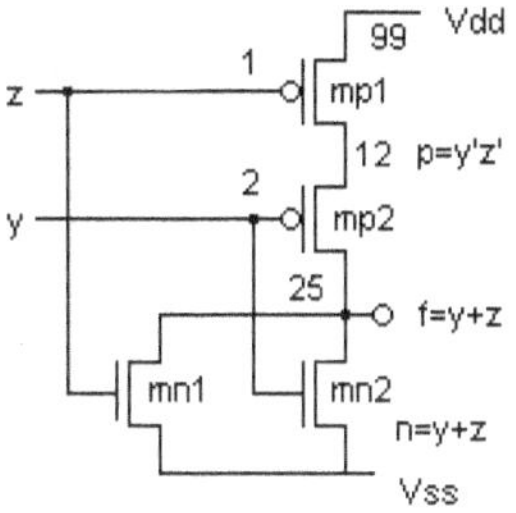

Click *List*. Double click *Frame300*. Add 1 pmos *M2V24x2* and 1 nmos *M2V5x2* to Frame300. Make the frame wider. Observe 3λ spacing from outline to transistor active, and from V_{SS}, V_{DD} active to transistor active. *Save* as *Frame304*.

Click *Text*. Add labels mp₁, mp₂, mn₁, mn₂. Complete the NOR layout in *Frame304* (1.3.7 page 27).

Figure 304e M2V5x2 (dots 1λ)

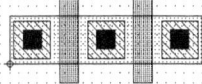

Figure 304g
Frame 304 (dots 2λ)

Figure 304f M2V24x2 (dots 1λ)

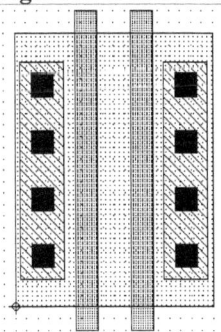

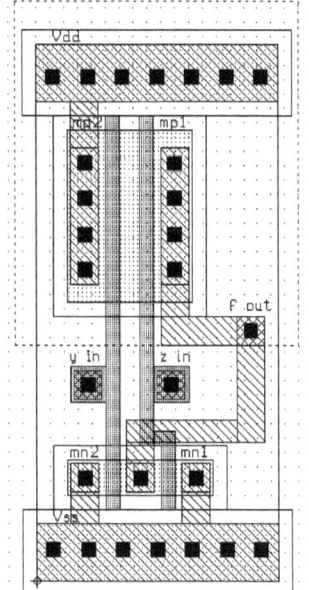

3.4 XOR *TG*

The 2-input xor *voltage* truth table shows the relationship of the inputs x, y, and output f_{xor}. Implementing XOR with transmission gates produces a simplified design (*the standard XOR circuit is 86*). The xor and xnor equations are

x	y	f_{xor}
L	L	L
L	H	H
H	L	H
H	H	L

(13a) $xor = x \oplus y = xy' + x'y$

(13b) $xnor = \overline{x \oplus y} = (x'+y)(x+y') = xy + x'y'$

TG XOR This preferred XOR circuit uses TG's .

Figure 306 *TG* XOR - 10 MOS, 2^+ Delays

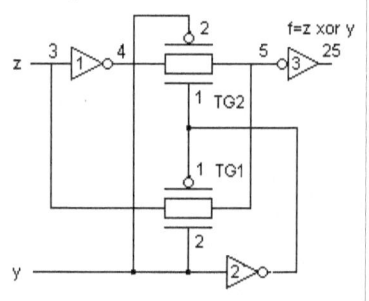

If y=H, then TG_2 is off and TG_1 is on so that the output V_{25} is f=yz' (z is inverted by inverter 3).

If y=L, then TG_1 is off and TG_2 is on so that the output V_{25} is f=y'z

The performance of a *minimum area* NOR with minimum nmos $W_n=5\lambda$ was satisfactory (equation 10c page 67, equation 12 page 68). Continue with a minimum area XOR, because we desire a minimum area chip. We increase C_{LOAD} to 40pf for a fanout=10 and double the inverter₃ W's.

Figure 30611 XOR *TG*, Cload/Cin=40fF/4fF

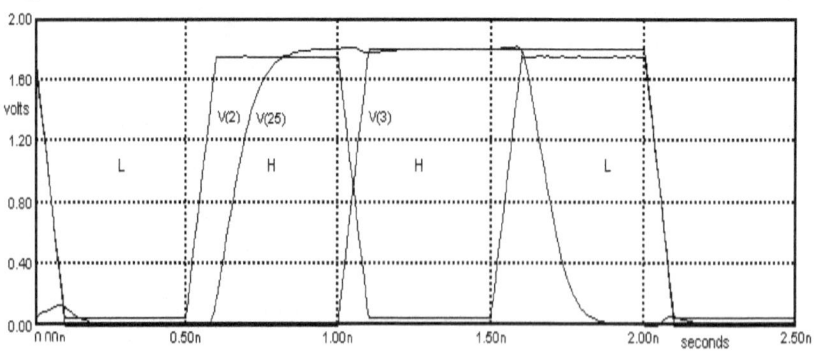

Problem 305 Reference Figures 306, 306h. Draw a Lasi layout for the 2-input XOR where the W's are as shown in Spice program 3061.

Spice program 3061

```
Fig3061.ckt  2-in xor Transient Response
.subckt mn1 103 102 101 99
MN1  103 102 101 99 N1  L=0.18u  W=0.45u          ; 5/2=W/L
+ AD=0.228p AS=0.228p PD=0.284u PS=0.284u
.ends mn1
.subckt mp3 209 208 207 298
MP3  209 208 207 298 P1  L=0.18u  W=1.08u          ; 12/2=W/L
+ AD=0.535p  AS=0.535p   PD=0.284u   PS=0.284u
.ends mp3
.include 180_N1P1.txt
Vdd 9  0  DC 1.8
V3   3 0 PULSE(1.80 0.00 0p 100p 100p 900p 2000p)  :z input
V2   2 0 PULSE(1.75 0.05 0p 100p 100p 400p 1000p)  :y input

xMP1    4   3  9  9 mp3        ; inv 1    Z input 12/2=W/L
xMN1    4   3  0  0 mn1                       5/2
xMP2    1   2  9  9 mp3        ; inv 2
xMN2    1   2  0  0 mn1
xMP3a  25  5  9  9 mp3        ; inv 3    W/L=24/2 f output
xMP3b  25  5  9  9 mp3
xMN3a  25  5  0  0 mn1                   W/L=10/2 f output
xMN3b  25  5  0  0 mn1
xMP4    5   1  3  9 mp3        ; TG 1
xMN4    5   2  3  0 mn1
xMP5    5   2  4  9 mp3        ; TG 2
xMN5    5   1  4  0 mn1
C25 25 0 40f    ; Cin=4fF
*.PLOT TRAN I(V1) I(V2) -500U,500U
*.PLOT TRAN V(2) V(2) V(3) V(4) V(5) V(25) 0,2
.TRAN 1e-012 2.5e-009 0 1e-011 UIC
.TEMP 27
.PLOT TRAN V(2) V(3) V(25) 0,2
.PRINT TRAN V(5)
.end
```

XOR Circuit Layout Process

The TG XOR circuit has 5 pmos and 5 nmos. The 5 nmos are placed in a p-well (anywhere on the p-wafer). The 5 pmos are placed in an n-well (that you create). The pmos are placed in a row whose area is intended to be

Figure 306b TG XOR

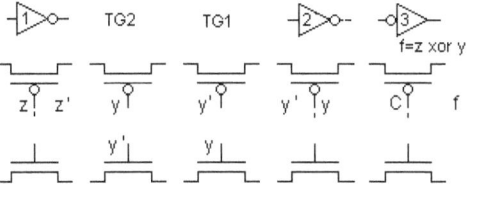

defined as one n-well. The nmos are placed in a row below the pmos (Figure 306b).

Figure 306c TG XOR with poly and metal₁

Add poly and metal₁ lines to define the pmos-nmos pairs (Figure 306c). Add metal₁ lines to define TG₁ and TG₂.

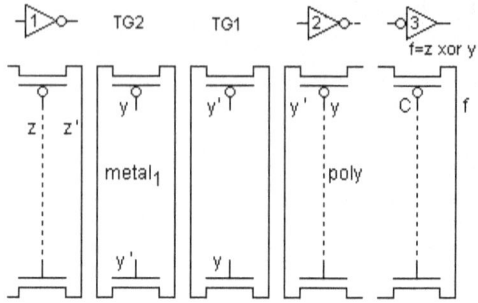

Poly is used to cross over metal₁, because metal₁ crossing metal₁ creates a short. That is why poly to metal₁ connections are made by vias that are plated holes in the oxide separating the poly and metal₁ layers. Small boxes designate vias connecting poly to metal₁. Poly and metal₁ lines are added as specified by the TG XOR circuit schematic (Figures 306, 306d). E.g. the 5 wire (Figure 306d) uses poly to cross over metal, and metal to cross over poly.

The transistor symbols are replaced by active boxes with vias and a poly gate (Figure 306e). The metal wires at TG1 and TG2 outputs are merged (node 5). The n-well and p select boxes enclosing the 5 pmos, the n-select boxes enclosing the 5 nmos are shown in the layout (Figure 306h).

Figure 306d TG XOR with poly and metal₁

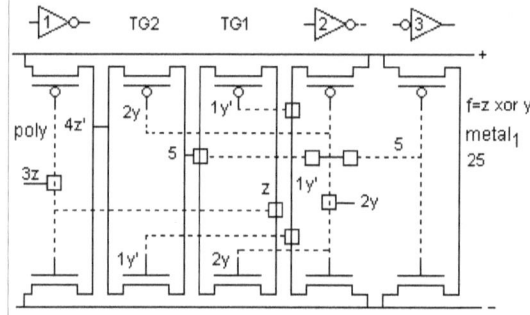

Figure 306e TG XOR

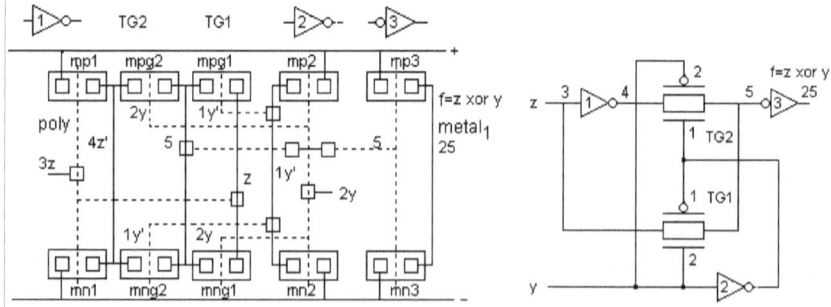

Figure 306f M3V5x2 (dots 1λ)

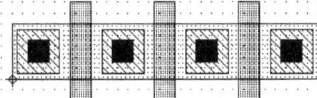

Figure 306g M3V12x2 (dots 1λ)

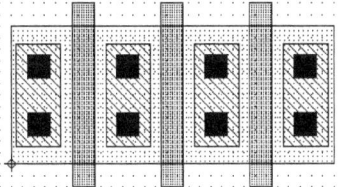

Figure 306h Frame306 (dots 2λ)

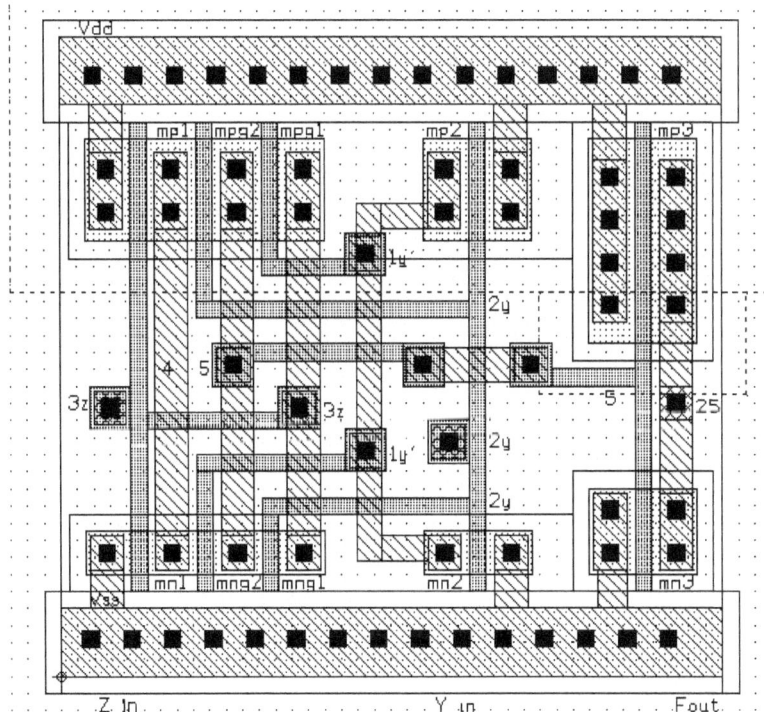

3.5 Multiplexer 4 To 1 *TG*

A mux is an OR of ANDs logic circuit, because AND represents selection of one input, and OR has the ability to route any selected input to the output. NOT gates are used to form the complements of address variables.

How the mux works Mux inputs and the output are active high variables. The two address lines y, z represent two binary digits (bits). Two bits can represent four numbers 0, 1, 2, and 3. This is a 36 transistor static circuit (Figure 307) consisting of four 6 transistor ANDS, one 8 transistor OR, and two 2 transistor NOTs.

Figure 307 four to one mux

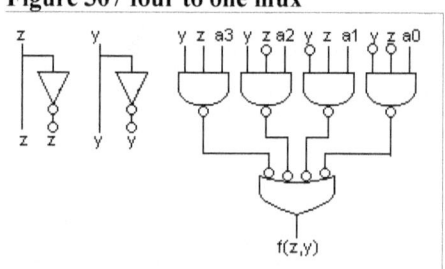

Our TG circuit design uses 15 transistors. (*The std 4:1 Mux circuit is 153*).

Address	y	z	$y'z'$	$y'z$	yz'	yz	f
0	0	0	a_0	0	0	0	a_0
1	0	1	0	a_1	0	0	a_1
2	1	0	0	0	a_2	0	a_2
3	1	1	0	0	0	a_3	a_3

The NOT circuits produce active high complements y' and z'. Each 3 input AND has one a_k input, one y or y' input, and one z or z' input. For example, truth table row 2 requires y,z values of 1,0 so that y, z' have values 1, 1 and the AND with inputs yz' (minterm m_2) is asserted. The world says the two bits y, z are decoded (Figure 308). (The address signals do not have to be periodic.)

Figure 308 4-line to 1-line multiplexer

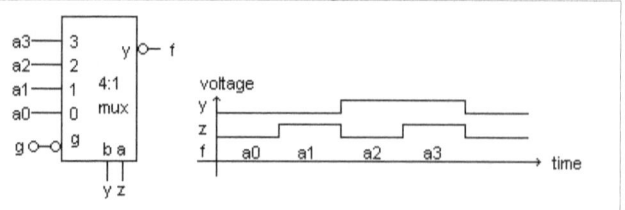

$g = 1$

y	z	f_{mux}
0	0	a_0
0	1	a_1
1	0	a_2
1	1	a_3

Transmission Gate
Inputs y and z are address lines, because they select a path, from one input to the output. Two bits have four states. This TG circuit has a low true output (Figure 309)

Figure 309 Multiplexer 4:1 *TG* , 15 transistors

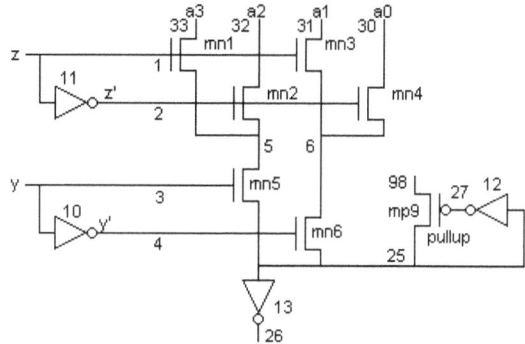

(14) $f_{mux} = a_0 y'z' + a_1 y'z$
$\qquad + a_2 yz' + a_3 yz$

y input	z input	path to node 25	output at node 26
L	L	a_0 mn4 mn6	a_0 (true when L)
L	H	a_1 mn3 mn6	a_1
H	L	a_2 mn2 mn5	a_2
H	H	a_3 mn1 mn5	a_3

Figure 30911 Multiplexer 4:1 TG signals z=V_1, y=V_3, f=V_{26}

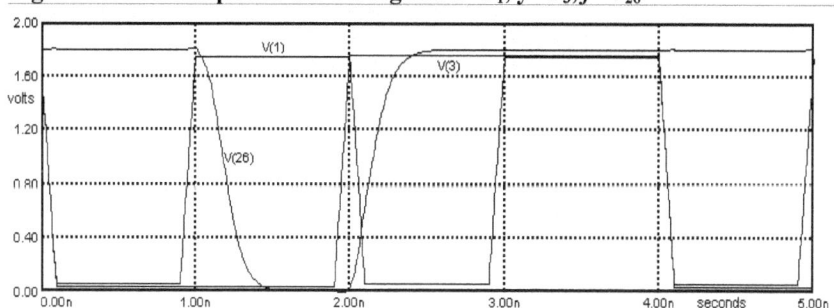

Figure 30912 Mux 4:1 TG V_{25} TG output, V_{27} pull-up, V_{26} output

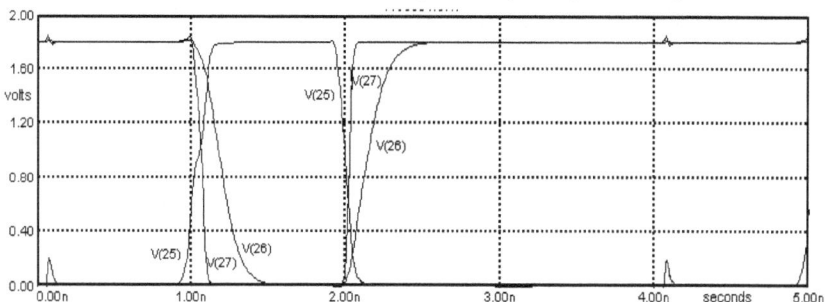

CMOS Circuit Design

TG Mux design We continue with use of minimum sized transistors.

TG Mux transient response In Spice program 3091 waveforms V_1 and V_3 produce yz values 0 to 3. The three inputs a_0, a_2, a_3 are at L, and a_1 is at H in order to produce an L output only when the address is LH=01_2 (Figures 30911 and 30912).

Spice program 3091 Mux 4:1
Fig3091.ckt Mux 4:1 TG Transient Response   * Level 8 (BSIM) SPICE models
.include 180_N1P1.txt

```
Vdd 9  0  DC 1.8

V1 1 0 PULSE(1.74  0.06  000p 100p 100p  800p 2000p)
V3 3 0 PULSE(1.76  0.04  000p 100p 100p 1800p 4000p)

*inverter 11
mp11   2  1  9  9 P1 L=0.18u W=1.08u        ;12/2
mn11   2  1  0  0 N1 L=0.18u W=0.45u        ; 5/2
*inverter 10
mp10  4  3  9  9 P1 L=0.18u W=1.08u         ;12/2
mn10  4  3  0  0 N1 L=0.18u W=0.45u         ; 5/2
* TG
mn1    5  1  0  0  N1  L=0.18u  W=1.08u     ;a3=L   12/2
mn2    5  2  0  0  N1  L=0.18u  W=1.08u     ;a2=L   12/2
mn3    6  1  9  0  N1  L=0.18u  W=1.08u     ;a1=H   12/2
mn4    6  2  0  0  N1  L=0.18u  W=1.08u     ;a0=L   12/2
mn5  25  3  5  0  N1  L=0.18u  W=1.08u      ;12/2
mn6  25  4  6  0  N1  L=0.18u  W=1.08u      ;12/2
*inv and pmos pullup
mp12   27 25  9  9 P1 L=0.18u  W=1.08u       ;12/2
mn12   27 25  0  0 N1 L=0.18u  W=0.45u       ;5/2
mp9     25 27  9  9 P1 L=0.18u  W=1.08u
* output inverter
mp13  26 25  9  9 P1  L=0.18u  W=1.08u       ;12/2
mn13  26 25  0  0 N1  L=0.18u  W=0.45u       ;5/2
C26   26  0 40f

.IC V(25)=0  V(27)=1.8
*.PLOT TRAN V(25) V(26) V(27) 0,2
.TRAN 1e-011 5e-009 0 1e-011
.TEMP 27
.PLOT TRAN V(1) V(3) V(26) 0,2
.end
```

MUX circuit Layout Process

Figure 309b MUX

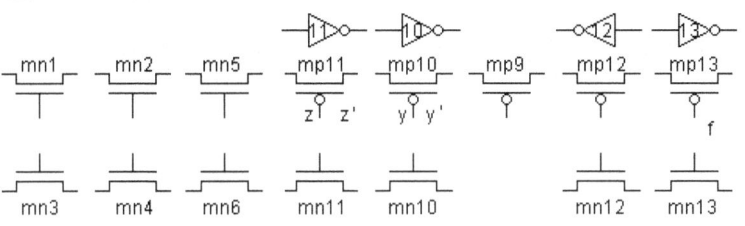

Figure 309c MUX

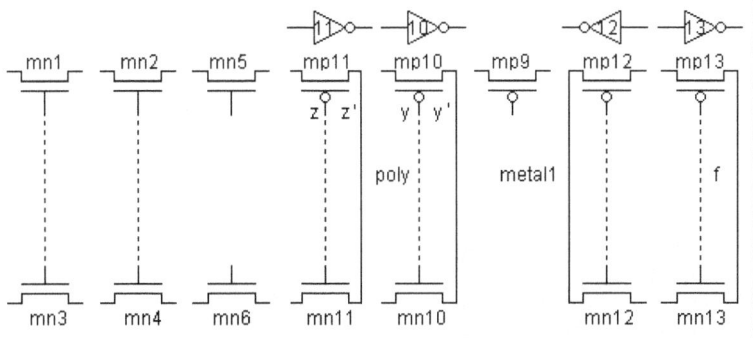

Figure 309d MUX

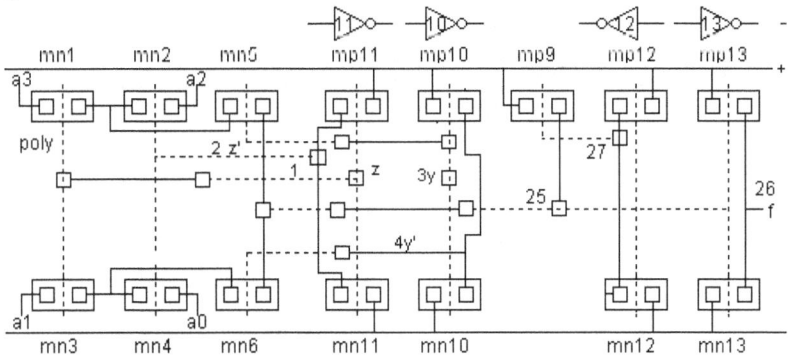

Note: the a_k inputs and output are only shown with labels a_0, a_1, a_2, a_3, f

Problem 306 Reference Figures 309d, 309g. Draw a Lasi layout for the 4:1 mux where the W's are as shown in Spice program 3091.

Figure 309g Frame309 (dots 2λ) Top to Bottom

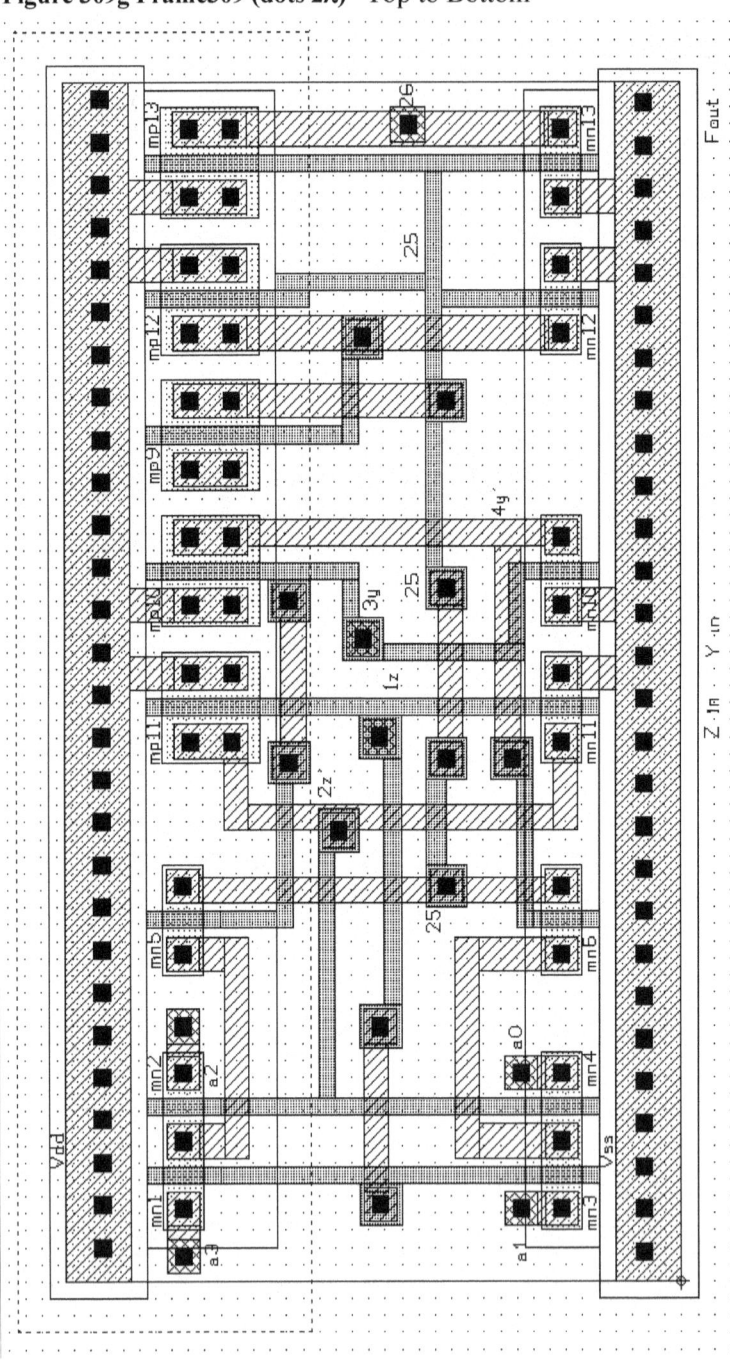

3.6 Full Adder Carry Circuit *Static*

The full adder carry circuit provides another complex example. For a positive true output c is in the pmos branch (equation 16a), and c' is in the nmos branch (equation 16b, Figure 310).

Full – adder :

Row	c_{in}	x	y	s	c
0	0	0	0	0	0
1	0	0	1	1	0
2	0	1	0	1	0
3	0	1	1	0	1
4	1	0	0	1	0
5	1	0	1	0	1
6	1	1	0	0	1
7	1	1	1	1	1

The mp$_1$ and mp$_2$ branch logic function is xy, because the active low x and y inputs turn on the pmos transistors. The mp$_4$ and mp$_5$ logic function is x+y, and mp$_3$ in series results in $c_{in}(x+y)$. So we get p=xy+c$_{in}$(x+y) for the p branch.

On the other hand the mn$_1$ and mn$_2$ branch logic function is x'y', because the active low x and y inputs turn off the nmos transistors. Series mn$_1$, mn$_2$ in parallel with mn$_3$

(15) $\quad s = x \oplus y \oplus c_{in}$

(16a) $\quad c = xy + c_{in}(x+y)$

(16b) $\quad c' = (x'+y')(c'_{in} + x'y')$

produces c$_{in}$'+x'y'. Mn$_5$ and mn$_4$ in parallel produce x'+y'. Thus the lower branch n logic is n=(x'+y')(c$_{in}$'+x'y'). When the upper branch is true f=H, and when the lower branch is true f=L.

Figure 310 Full adder carry circuit Static   **Figure 311 Transistor Widths**

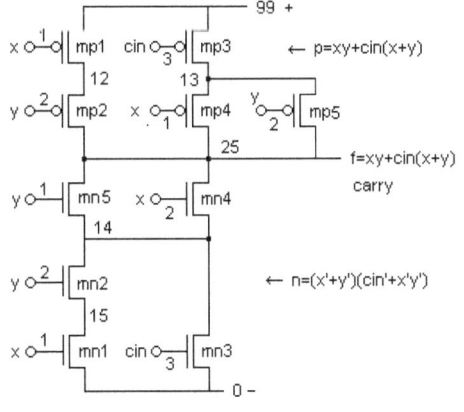
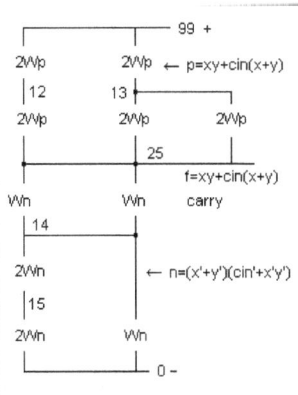

Problem 307 Reference Figures 310, 310e. Draw a Lasi layout for transistors M2V10x2, M2V12x2, M1V05x2, and the Full Adder Carry circuit with the W's used in Spice program 3101.

Spice program 3101

Fig3101.ckt full adder carry static

```
.include 180_N1P1.txt
Vdd 99  0  DC 1.8

V1   1 0 PULSE(1.75 0.05 0p 100p 100p  400p 1000p)        ; x
V2   2 0 PULSE(1.80 0.00 0p 100p 100p  900p 2000p)        ; y
V3   3 0 PULSE(1.70 0.05 0p 100p 100p 1900p 4000p)        ; cin

V4   4 0 PULSE(0.05 1.75 0p 100p 100p 400p 1000p)

MP1 12  1 99 99 P1  L=0.18u  W=2.16u             ; 24/2=W/L
MP2 25  2 12 99 P1  L=0.18u  W=2.16u
MP3 13  3 99 99 P1  L=0.18u  W=2.16u
MP4 25  1 13 99 P1  L=0.18u  W=2.16u
MP5 25 2 13 99 P1  L=0.18u  W=2.16u

MN1  15  1  0  0 N1  L=0.18u  W=0.90u             ; 10/2
MN2  14  2 15  0 N1  L=0.18u  W=0.90u
MN3  14  3  0  0 N1  L=0.18u  W=0.45u             ; 5/2
MN4  25  1 14  0 N1  L=0.18u  W=0.45u
MN5  25  2 14  0 N1  L=0.18u  W=0.45u

C25 25 0 20f        ; Wgate=2λ/1.4 x Cload= 2λ/1.4 x 20=28.6 λ, υσε 24 λ

.TRAN 1e-012 5e-009 0 1e-011 UIC
.TEMP 27
.PLOT TRAN V(1) V(25) 0,2
.PLOT TRAN V(2) V(3) 0,2
.PRINT TRAN V(3)
.end
```

Figure 31011 Full Adder Carry output HHHLHLLL (truth table rows 0 to 7)

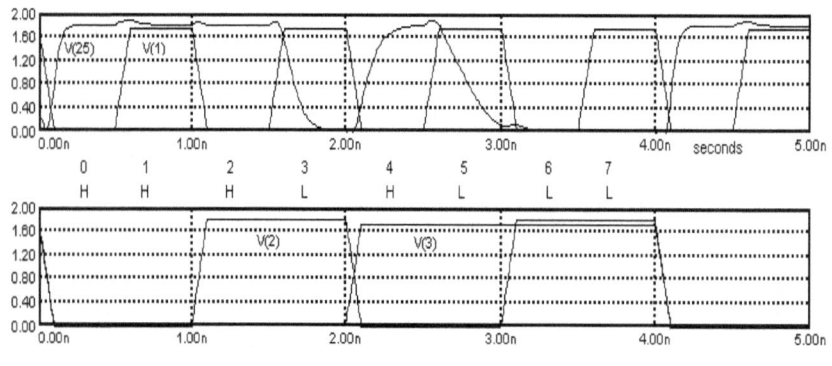

Figure 310b Carry Circuit Transistors

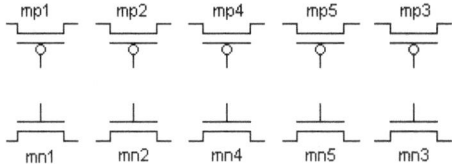

Figure 310c

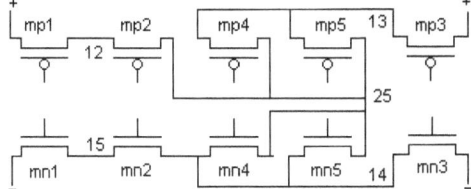

Figure 310d

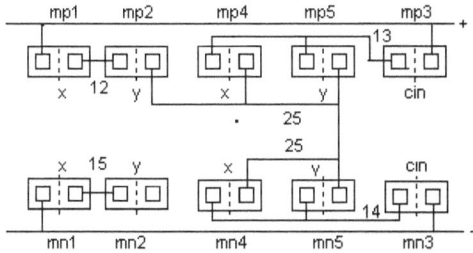

Figure 310e Carry Circuit (dots 2λ)

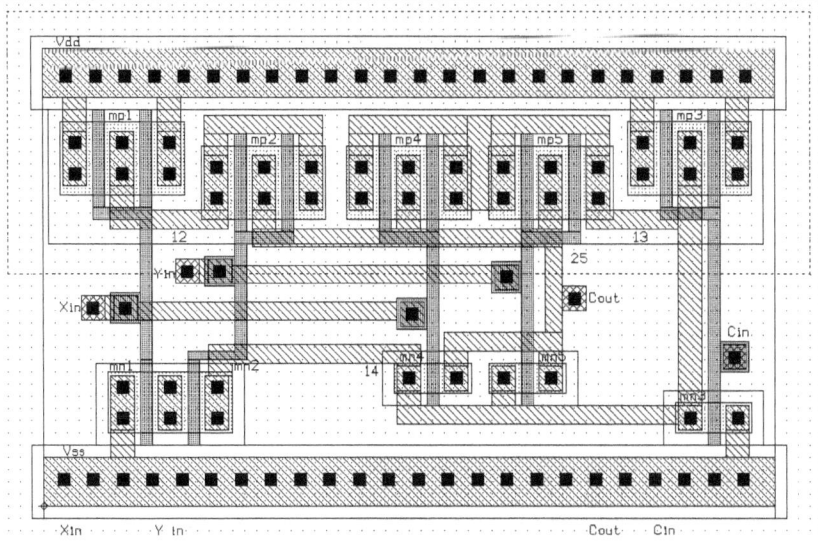

4 One Bit Digital Circuits with Memory

A one-bit synchronous bistable memory circuit is called a flip-flop, whereas a one-bit asynchronous bistable memory circuit is called a latch. An asynchronous logic circuit is converted to a synchronous logic circuit when you add components so that circuit outputs are changed only by active clock edges. At the clock input if an active clock edge flips an output from L to H, then a subsequent active clock edge can flop it back to L. The active clock edge can positive *or* negative going.

Design is an cyclic process, because adding transistors adds capacitance. Additional capacitance requires transistor W changes. The number of iterations can be minimized if we estimate this additional capacitance. We make a conservative assumption that node C_{load} is 20fF. This becomes the basis for initial sizes of transistors driving internal nodes in many circuits.

Note on Power dissipation The power used by a circuit is $V_{dd} \times I_{avg}$. Current is drawn from V_{dd} when pmos transistors turn on, because the capacitors at the pmos drains are charged to V_{dd}. When the nmos transistors turn on the same capacitors are discharged to zero volts thereby returning the charge to the supply. CMOS circuits only draw current during transitions. That is why power is proportional to the number of transitions per second. There are other sources of power dissipation such as leakage currents that are much smaller. Smaller perhaps, but significant when a system is "off" and battery life is an issue.

charge transferred $dq = idt \Rightarrow Q_c = I_{avg} T_{clock\,period}$

also $Q_c = C_{system} V_{dd} \Rightarrow C_{system} V_{dd} = I_{avg} T_{clock\,period} = I_{avg} \dfrac{1}{f_{clock}}$

hence $I_{avg} = C_{system} V_{dd} f_{clock}$

the power P_{avg} *dissipated is* $P_{avg} = I_{avg} V_{dd} = C_{system} V_{dd}^2 f_{clock}$

4.1 SR Latch *Static*

A logic circuit acquires the memory property when a circuit output is connected to a circuit input. This connection from an output to an input is known as feedback. The most elementary logic circuit with the memory property is an SR Latch (set-reset latch) assembled from two gates. Starting with one gate a pulse at the input is inverted and replicated at the output after a time delay t_d known as a propagation delay (Figure 401a). Propagation delay is the time required for changes at an input to appear at an output. This SR latch assembled from NANDs is usually drawn as in Figure 401c. The mixed logic schematic (Figure 401b) shows circuit actions clearly when compared to Figure 401c.

When output q follows input s, the circuit is *transparent*. The circuit is no longer transparent when feedback is added by connecting an output signal q to an AND input. (Figure 401b).

Figure 401 SR Latch using nand gates

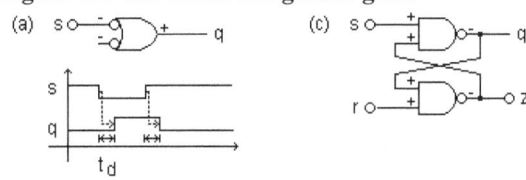

Assume s and r are H. Also assume q is L so that z is H and the latch is in the reset state.

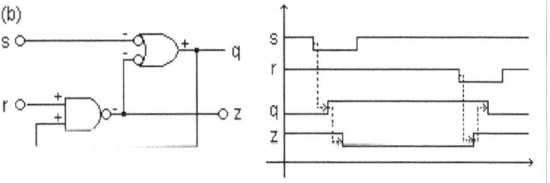

Set When s asserts and goes L, q is driven H (Figure 401b). Now, with both r and q at H, z is driven L to hold q at H. Later s returns to H, but q remains at H, because z is L. The latch is now set.

Figure 402 Static SR Latch - two nand gates

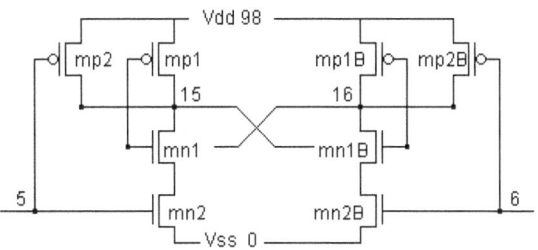

Reset Later, when r asserts and goes L, z is driven H. Now, with both s and z at H, q is driven L to hold z at H. Later r returns to H, but q remains at L, because s is H and z is still H. The latch is now reset.

CMOS Circuit Design

Design We need to size the latch's transistors. Assume external C_{load}=20fF.

transistor drive current In Section 2.3 page 45 normalized values for mos drain currents per λ and μm were calculated.

(1) $i_{ds} = i_{drive} \dfrac{mA}{\mu m} \times W \; \mu m$ where $i_{ndrive} = 0.61 \dfrac{mA}{\mu m}$ $i_{pdrive} = 0.26 \dfrac{mA}{\mu m}$

Since drain current charges and discharges the load capacitance while switching we can calculate W from the capacitor vi constraint.

(2) $W = \dfrac{i_{ds}}{i_{drive}} \mu m = \dfrac{1}{i_{drive}} C_{load} \dfrac{dv}{dt}$ $\Rightarrow$ $W = \dfrac{C_{load}}{i_{drive}} \dfrac{dv}{dt}$

Assume C_{load} = 20fF at nodes 15 and 16 (Figure 402). The NAND pmos and nmos transistors are sized as follows.

(3a) $W_n = \dfrac{C_{total}}{i_{ndrive}} \dfrac{dv}{dt} = \dfrac{20\,fF}{0.61mA/\mu m} \cdot \dfrac{1.8}{200\,ps} = 0.295\mu m = 3.27\lambda \approx 5\lambda$

(3b) $W_p = \dfrac{i_{ndrive}}{i_{pdrive}} W_n = \dfrac{0.61mA}{0.29mA} \cdot 5\lambda = 10.5\lambda \approx 12\lambda$

The nmos are in series, so double W_n to 10λ.

Here is the C_{gate} calculation.

(4a) $C_{gate} = C_{ox}WL = C_{ox}\dfrac{W}{L}L^2 = \left(C_{ox}L^2\right)\dfrac{W}{L}$

(4b) $C_{gate} = 8.63 \dfrac{fF}{\mu m^2}(0.18\mu m)^2 \dfrac{W}{L_{min}} = 0.280 \dfrac{W}{2\lambda} fF$ $(L_{min} = 2\lambda = 0.18\mu m)$

(5) $C_{gate} = 0.280 \dfrac{W_p + W_n}{2\lambda} fF = 0.280 \dfrac{(12+10)\lambda}{2\lambda} = 3.08\,fF$

Figure 40211 Static SR Latch Transient Response

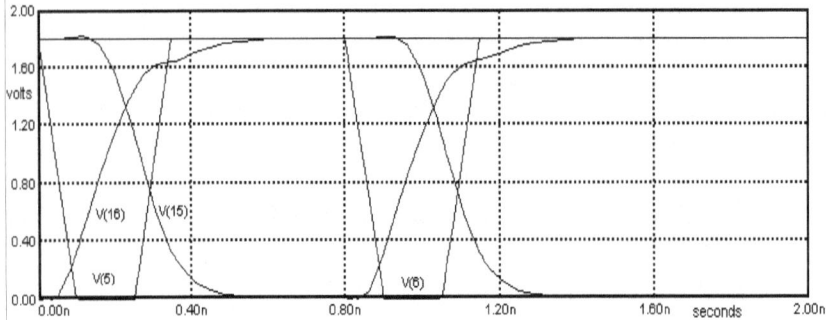

Spice program 4021
Fig4021.ckt SR latch Transient Response

* Level 8 (BSIM) SPICE models
.include 180_N1P1.txt

Vdd 98 0 DC 1.8

V6 6 0 PULSE(1.8 0 800p 100p 100p 150p 2000p)
V5 5 0 PULSE(1.8 0 000p 100p 100p 150p 2000p)

.subckt mp3 209 208 207 298
MP3 209 208 207 298 P1 L=0.18u W=1.08u   ; 12/2=W/L
+ AD=0.535p AS=0.535p PD=3.15u   PS=3.15u
.ends mp3

.subckt mn2 118 117 116 115
MN2 118 117 116 000 N1 L=0.18u W=0.90u   ;10/2=W/L
+ AD=0.446p AS=0.446p PD=2.79u PS=2.79u
.ends mn2

.subckt nand1 1 2 25 98
XMP1 25 1 98 98 mp3   ;12/2=W/L
XMP2 25 2 98 98 mp3
XMN1 25 1 13   0 mn2   ;10/2
XMN2 13 2   0   0 mn2
.ends nand1

*latch
Xnand1 5 16 15 98 nand1
Xnand2 6 15 16 98 nand1
C15 15 0 20f   IC=0
C16 16 0 20f   IC=1.8

.TRAN 1e-011 2e-009 0
.TEMP 27
.PLOT TRAN V(5) V(6) V(15) V(16) 0,2
.END

Problem 401 Draw a layout for transistors M1V12x2, M1V10x2, and the Latch using W's from spice program 4021.

Figure 402b Latch

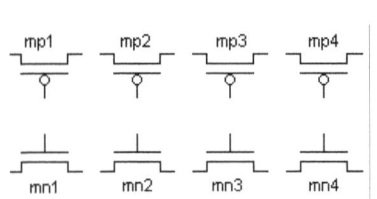

Figure 402c Latch

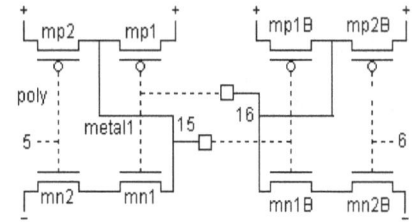

Figure 402d Latch

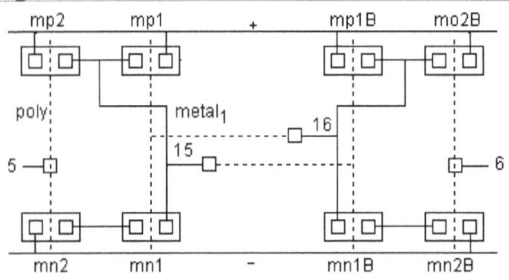

Metal$_2$ wires not shown.

Figure 402e Latch (dots 2λ)

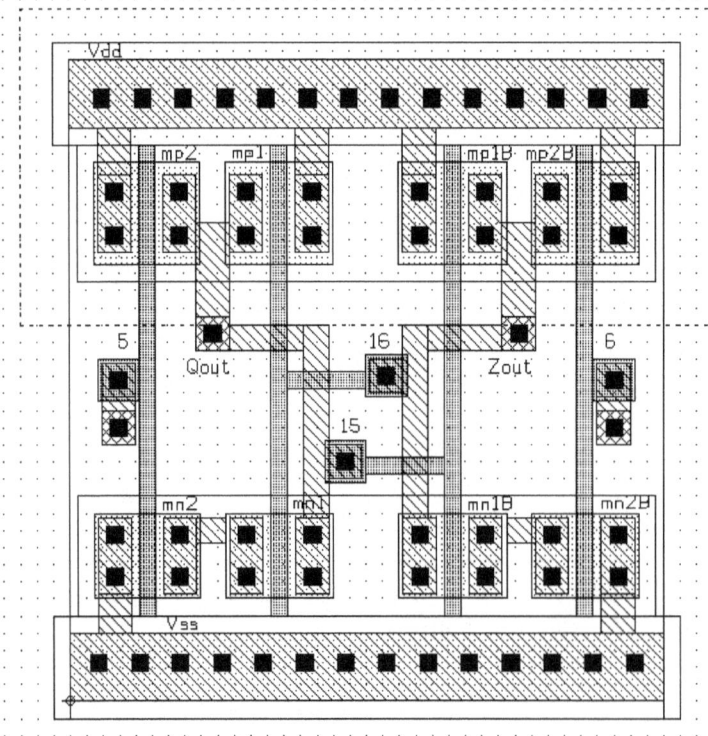

4.2 Flip-Flop *TG*

Transparent when clock is L The flip-flop is *transparent* when clock c=L and c_{bubble}=H, because then TG_{11} is closed and TG_{12} is open (Figure 403c). Transparent means output q equals input d after propagation time elapses.

When clock c transitions from L to H, TG_{11} opens and TG_{12} closes. Input d is disconnected, and the value of q is *latched* a long as clock is H (Figure 403b). Note that input d must be stable before the L to H clock transition.

When the clock transitions from H to L, TG_{11} closes again and TG_{12} opens to reenter the transparent state. Input d is reconnected, and the value of q changes to the new value of d. And so forth.

Figure 403 Flip-flop transparent **on clock L, and latched on clock H**

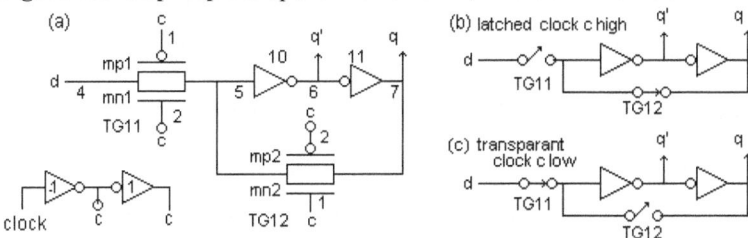

Transparent when clock is H If we exchange the clock connections to the TG's, then the flip-flop is transparent when clock is H, and latched when clock is L (Figure 404).

Figure 404 Flip-flop transparent **on clock H, and latched on clock L**

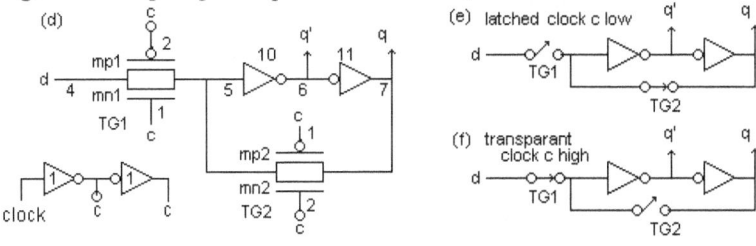

Problem 402 Draw a layout for transistor M1V05x2 and the Flip-flop in Figure 403 where W_p=12λ and W_n=5λ.

Clock A clock is an independent signal source that emits a periodic signal. In a digital system the clock waveform is essentially a squarewave (Figure 498a). The transition from L to H is referred to as the positive edge (shown as up arrows in Figure 498c), and the transition from H to L is referred to as the negative edge (shown as down arrows in Figure 498d). The transition times are referred to as the rise and fall times respectively. Idealized waveforms with zero transition times make timing diagrams easier to perceive (Figure 498b).

Figure 498 Periodic Clock Waveform

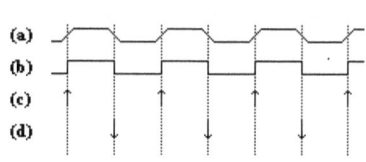

Periodic Clock Waveform The time between two consecutive positive, or two consecutive negative, transitions is the clock's *period* that is the reciprocal of the clock's *frequency*. A 500MHz clock has a 2ns period. We emphasize that there is one *triggering edge* per period by showing the clock waveform as a series of up arrows (Figure 498c), or down arrows (Figure 498d).

Aperiodic edges For example, the periodic clock can be "shut off." Then a special circuit allows one pulse, with one triggering edge, to be emitted by one press of a push button.

The Setup and Hold Window
Synchronous circuits have a clock signal input, and constraints on the data inputs. 1) Inputs cannot change during a clock transition, and 2) the circuit must be in a stable condition during a clock transition. The

Figure 499 Setup and Hold Window

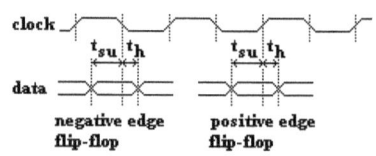

constraints are satisfied if there are no input changes inside *the setup and hold window* (Figure 499). Only the clock changes inside the window.

The setup time (t_{SU}) is a quiet time *before* an active clock edge. The setup time is the time interval before the clock transition during which no inputs should change (Figure 499). The hold time (t_H) is a quiet time *after* an active clock edge. The hold time is the time interval after the clock transition during which no inputs should change (Figure 499).

4.3 Edge Triggered Flip-Flops

At the end of a long design and development cycle a variety of flip-flop designs narrowed down to two commercially available *edge triggered* flip-flops: the D and JK flip-flops. A flip-flop has two states known as set and reset. The output's symbol is q. The output q reports the state of the flip-flop. That is why q is called a *state variable. q's next state is shown as q^+*. A flip-flop's next state defining equation is the most important fact a designer needs to know. Everything is derived from that equation.

Defining equations :

(6a) $d\ flip$ - $flop$: $q^+ = d$

(6b) $jk\ flip$ - $flop$: $q^+ = jq' + k'q$

Figure 405 D and JK flip-flops

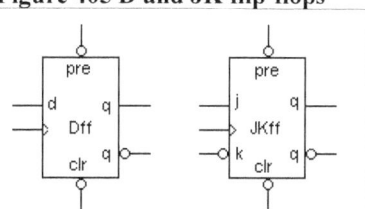

A *D* flip-flop has two synchronous inputs: a clock input and a *d* input for defining the next state (Figure 405). Asserting the asynchronous preset or clear inputs sets or resets the q outputs, while overriding the synchronous inputs. When the asynchronous preset or clear inputs are not asserted the *d* defining equation is operational. Then the present state data at the *d*-input meeting setup and hold requirements is transferred to the next state output q on the next positive clock edge. Triggering occurs when the edge rises up past a voltage level, and is not dependent on edge's rise time as a first approximation. After the hold time interval elapses data *d* may be changed (Figure 499).

A JK flip-flop has three synchronous inputs: a clock input and *j* and *k* inputs for defining the next state (Figure 405). Asserting the asynchronous preset or clear inputs sets or resets the q outputs, while overriding the synchronous inputs. When the asynchronous preset or clear inputs are not asserted the *jk* defining equation is operational. Then the present state data at the *jk* inputs meeting setup and hold requirements is transferred to the next state output q on the next positive clock edge. Triggering occurs when the edge rises up past a voltage level, and is not dependent on edge's rise time as a first approximation. After the hold time interval elapses data *jk* may be changed.

4.3.1 D Flip-Flop *Static and TG*

A d flip-flop has two output lines for q. One output is active-high and the other is active-low. A complete truth table for the D flip-flop includes inputs pre, clr, d, and the clock. However we emphasize the truth table for the d logical function.

d	f_d
0	0
1	1
$f_d = d$	

D flip-flop timing diagram If any clock period n is the current period then *n is the present period*, and clock period *n+1 is the next period*. The value of d in the present period is the *present state of d*. Since $q^+ = d$ the output in clock period n+1 is the *next state of q* that is a copy of present state d (Figure 406). Thus the q waveform is the same as the d waveform delayed in time by one clock period. A one-clock-period slip assumes the d input is synchronous with the clock. We do not show a q^+ waveform, because in the next state q^+ becomes the present-state q.

The coincidence of the d values and the clock edge are marked by circles (Fig. 406). The circles mark the cause and arrows show the effect. The propagation delays from clock to output must be greater than zero. When t_{PHL} and t_{PLH} are

Figure 406 D Flip-flop timing diagram

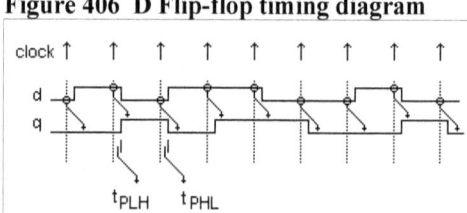

not greater than zero, then the d and q waveform transitions merge with the clock edge. Then equation $q^+ = d$ cannot answer the question "What is the value of d at the clock edge?" When propagation delays are zero there is no definite answer. This is why propagation delays greater than zero are necessary for synchronous operation. Since all physical circuits have propagation delays greater than zero this is not a problem.

An edge triggered Static D flip-flop A static D flip-flop circuit is an assembly of two latches plus glue logic gates for gating the clock and converting the d input into s and r latch inputs (Figure 407). The

Figure 407 D Flip-flop *Static*

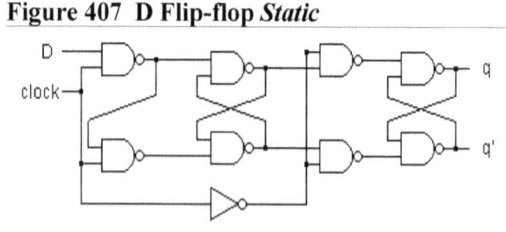

circuit uses 8×4+2=34 transistors without pre and clr inputs. On the other hand a TG circuit design has 20 transistors.

An edge triggered TG D Flip-flop This circuit is an assembly of inverters and TGs (Figure 408) that are two flip-flops in cascade (4.2 page 89). This is an edge triggered circuit, because the circuit output q changes state only when a clock edge goes from L to H.

clock low We start with clock c at L and c_{bubble} at H, so that the *master* is transparent (TG$_1$ on, TG$_2$ off, Figure 409b), and the *slave* is latched (TG$_3$ off, TG$_4$ on). The d value at node 3 is passed to transparent nodes 4, 5, and 6 (Fig 409b). The slave positive feedback loop (nodes 7, 8, 9) holds the value received from node 5 when c was H in the prior period.

clock high When clock c goes H the *master* latches, and the *slave* becomes transparent (Figure 409a). With TG$_1$ off and TG$_2$ on, the master positive feedback loop (nodes 4, 5, 6) holds the value received when c was L before going H. Since TG$_3$ is on, and TG$_4$ is off, the value at node 5 is passed to nodes 7, 8, 9 to become the next state of q (Figures 409a, 410).

clock low When clock c goes L the *master* becomes transparent, and the *slave* latches to hold the next state of q until clock goes H again. This ends one clock period.

Figure 408 Edge triggered D flip-flop

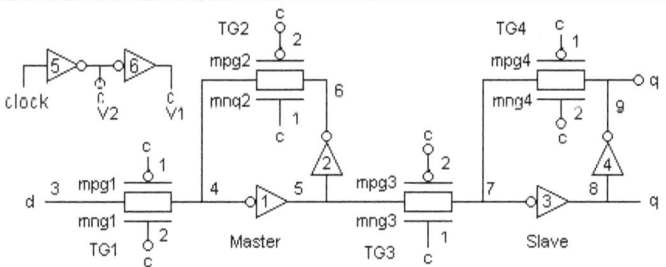

Figure 409 Edge triggered D flip-flop (a) clock high (b) clock low

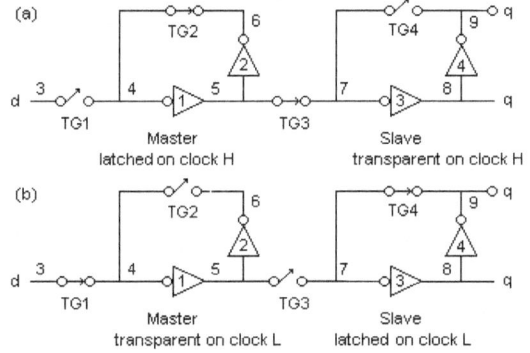

Figure 410 D flip-flop Waveforms

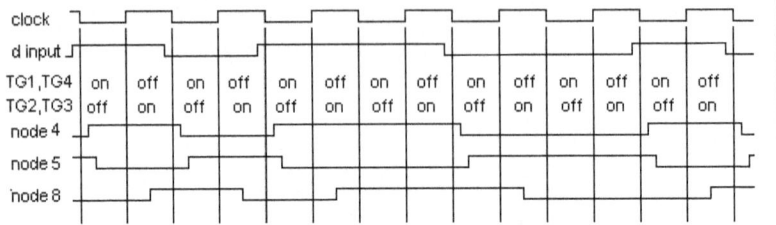

Design We start by sizing the transistors of inverters 3 and 4 assuming C_{load}=25fF at nodes 8 and 9 (Figure 408). The nmos "minimum" sizes are as follows (estimating C_{gate} and $C_{transistors}$).

$$(7a) \quad W_n = \frac{C_{load}}{i_{ndrive}} \frac{dv}{dt} = \frac{25\,fF}{0.61mA/\mu m} \cdot \frac{1.8}{200\,ps} = 0.369\mu m = 4.1\lambda \approx 5\lambda$$

$$(7b) \quad W_p = \frac{i_{ndrive}}{i_{pdrive}} W_n = \frac{0.61mA}{0.29mA} \cdot 4.1\lambda = 8.62\lambda \approx 12\lambda$$

$$(8) \quad C_{gate3/4} = 0.280 \frac{W_p + W_n}{2\lambda} fF = 0.280 \frac{(12+5)\lambda}{2\lambda} = 2.38\,fF$$

Use W_p=12λ, W_n=5λ, and C_{Load}=25fF for inverters 3, 4 in Spice program 4081. Fanout = 25/2.38 =10.5.

Use minimum sizes W_n=5λ, W_p=12λ for all TGs and inverters. Sizes can be increased if performance improvement is required. Figures 40812, 40813 show satisfactory performance.

Problem 403 Draw a layout for the Edge triggered D Flip-flop (Figures 408, 408e). Use W_p=12λ and W_n=5λ.

Problem 404 Draw a schematic showing preset and clear inputs. Hint change some inverters to 2-input NORs.

Figure 408 Edge triggered D flip-flop

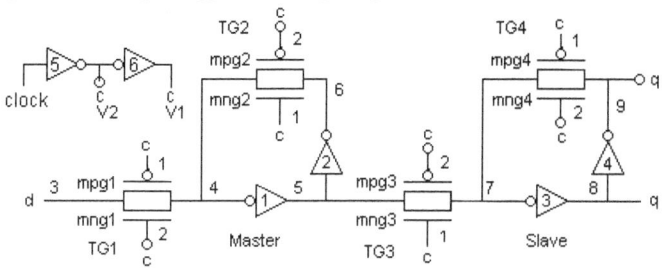

Figure 40811 D Flip-flop clock signals V_1, V_2

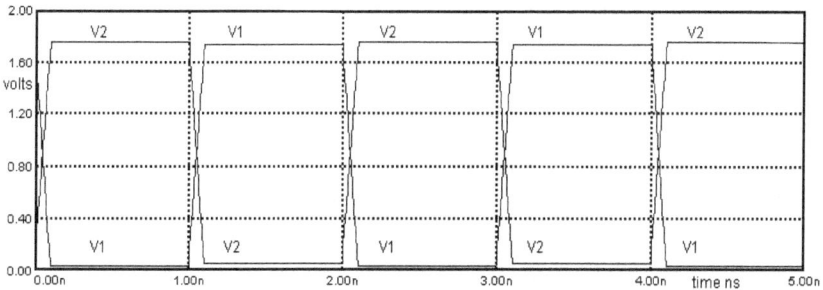

Figure 40812 D Flip-flop signals d =V_3, master V_4, V_5

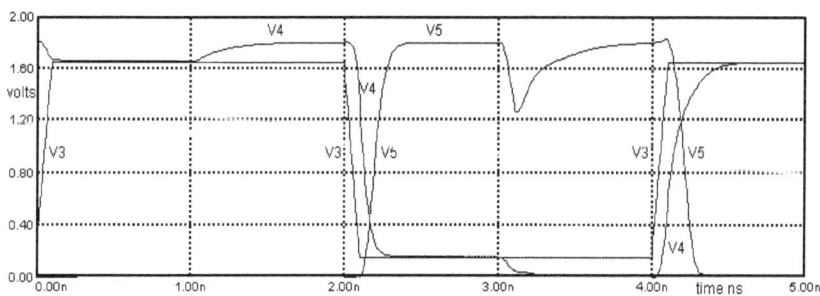

Figure 40813 D Flip-flop signals slave d=V_3, q output=V_8, qbar output = V_9

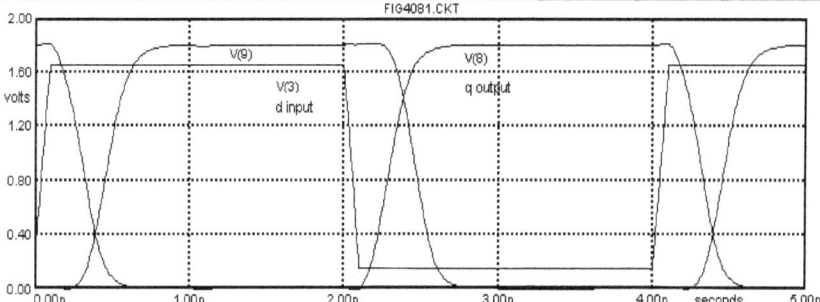

Spice Program 4081
Fig4081.ckt D flip-flop TG Transient Response
* Level 8 (BSIM) SPICE models
.include 180_N1P1.txt

Vdd 98 0 DC 1.8
V1 1 0 PULSE(0.06 1.76 000p 100p 100p 900p 2000p)
V2 2 0 PULSE(1.74 0.04 000p 100p 100p 900p 2000p)
V3 3 0 PULSE(0.15 1.65 000p 100p 100p 1900p 4000p)

.subckt mn1 103 102 101 99
MN1 103 102 101 99 N1 L=0.18u W=0.45u   ; 5/2=W/L
+ AD=0.223p AS=0.223p PD=1.89u PS=1.89u
.ends mn1
.subckt mp3 209 208 207 298
MP3 209 208 207 298 P1 L=0.18u W=1.08u   ; 12/2=W/L
+ AD=0.535p AS=0.535p PD=3.15u   PS=3.15u
.ends mp3
****
*TG1
xmp1 3 1 4 98 mp3 ;12/2
xmn1 3 2 4 0 mn1 ; 5/2
*inverter 1
xmp11 5 4 98 98 mp3
xmn11 5 4 0 0 mn1
*inverter 2
xmp12 6 5 98 98 mp3
xmn12 6 5 0 0 mn1
*TG2
xmp2 6 2 4 98 mp3
xmn2 6 1 4 0 mn1
****
*TG3
xmp3 5 2 7 98 mp3
xmn3 5 1 7 0 mn1
*inverter 3
xmp13 8 7 98 98 mp3
xmn13 8 7 0 0 mn1
*inverter 4
xmp24 9 8 98 98 mp3
xmn24 9 8 0 0 mn1
*TG4
xmp4 9 1 7 98 mp3
xmn4 9 2 7 0 mn1
****

.IC V(7)=0
C8 8 0 25f   ; fanout = 10
C9 9 0 25f

*Fig40811 .PLOT TRAN V(1) V(2) 0,2
*Fig40812 .PLOT TRAN V(3) V(4) V(5) 0,2
*Fig40813 .PLOT TRAN V(3) V(8) V(9) 0,2

.TRAN 1e-011 5e-009 0 1e-011
.TEMP 27
.PLOT TRAN V(3) V(8) V(9) 0,2
.end

Figure 408b D Flip-flop

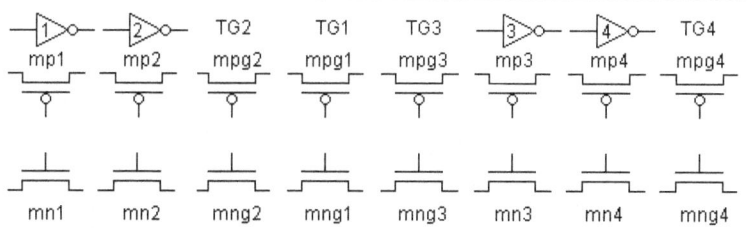

Figure 408c D Flip-flop

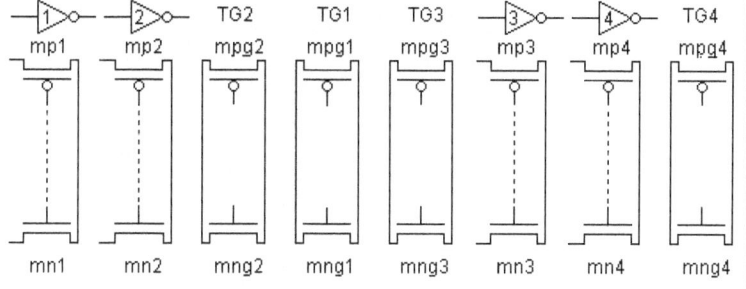

Figure 408d D Flip-flop (2 clock inverters omitted)

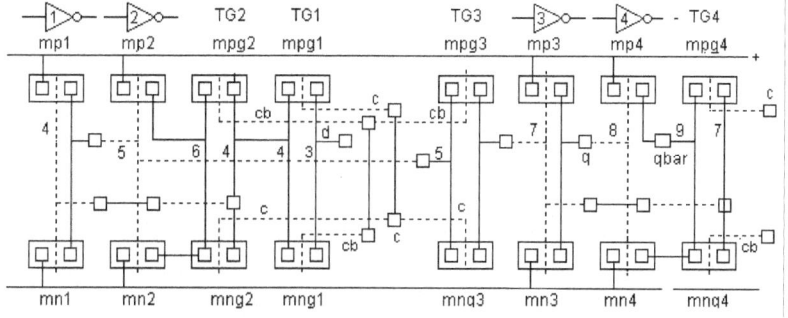

Figure 408e D Flip-flop (dots 2λ) (2 clock inverters omitted)

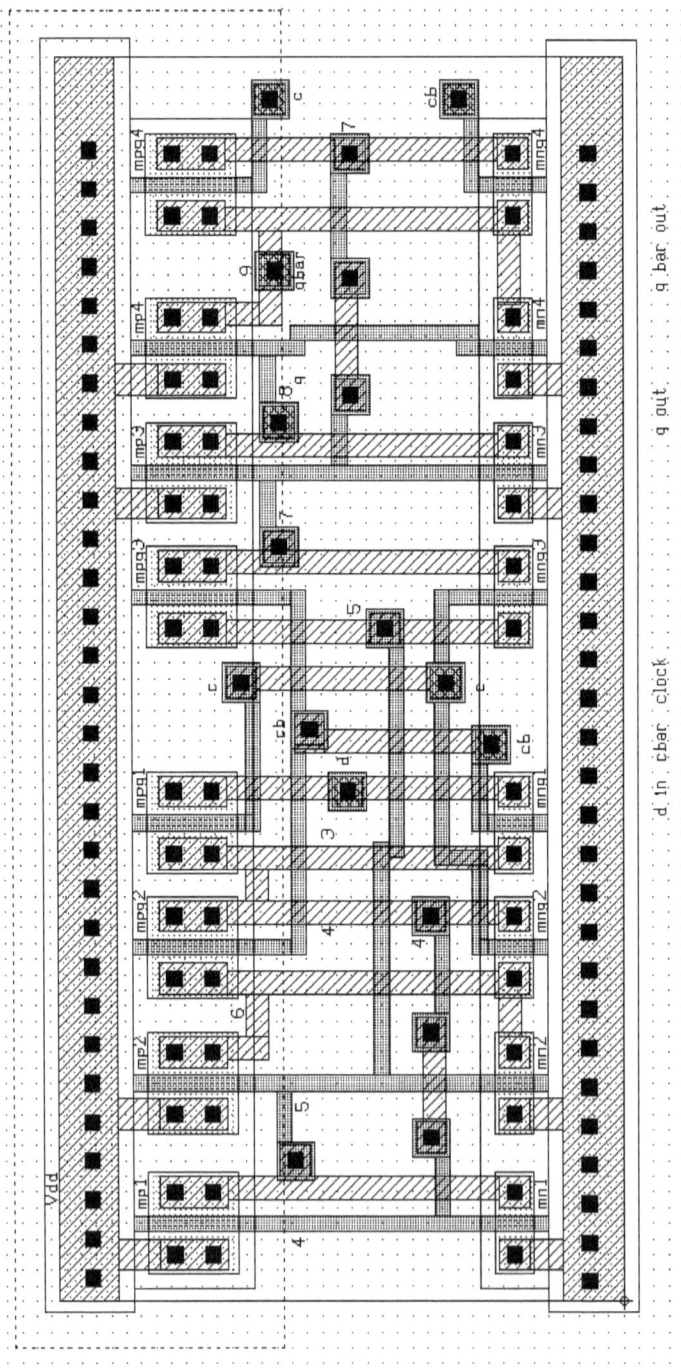

4.3.2 JK Flip-Flop *Static and TG*

The synchronous *jk* flip-flop has two output lines for *q*. One output is active-high and the other is active-low. A complete truth table for the *jk* flip-flop includes inputs pre, clr, j, k, and the clock. However we emphasize the truth table for the *jk* logical function.

j	k	f_{jk}	mode
0	0	q	hold
0	1	0	reset
1	0	1	set
1	1	q'	toggle

$$f_{jk} = qj'k' + jk' + q'jk$$

The j and k inputs define the next state. The (present state) data at the *jk* inputs meeting setup and hold requirements are transferred to the (next state) output q on the next positive clock edge.

(9) *jk flip - flop* : $q^+ = jq' + k'q$

This is the jk defining equation. If the present state q is zero then $f_{jk}=j$ so that k is a don't care input. The next state depends only on j. If the present state q is one then $f_{jk}=k'$ so that j is a don't care input. The next state only depends on k'. Then the *jk* excitation table above can be recast as a very useful state transition table (– is don't care).

q transition	j	k
0 to 0	0	–
0 to 1	1	–
1 to 0	–	1
1 to 1	–	0

An edge triggered Static jk flip-flop A static *jk* flip-flop circuit is an assembly of two latches plus glue logic gating the clock, and converting the j, k inputs into s and r latch inputs (Figure 411). The circuit

Figure 411 JK flip-flop *Static*

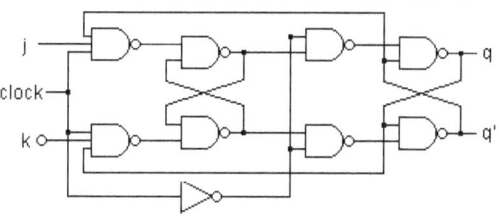

uses 6×4+2×6+2=38 transistors without pre and clr inputs (Figure 411). On the other hand a TG circuit design has 32 transistors.

An edge triggered TG JK flip-flop This circuit is an assembly of inverters and TGs (Figure 412) that are two d flip-flops in cascade (4.2 page 89) plus glue logic implementing d=jq′+k′q. This is an edge triggered circuit, because the latches change state (transparent to latch and vice versa) on clock edges (H to L and L to H).

(9) *jk flip - flop :* $q^+ = jq' + k'q$

clock low We start with clock c at L and c_{bubble} at H, so that the *master* is transparent (TG$_1$ on, TG$_2$ off), and the *slave* is latched (TG$_3$ off, TG$_4$ on). The value of d at node 3 is passed to transparent nodes 4, 5, and 6 (Figure 412). The slave positive feedback loop (nodes 7, 8, 9) holds the value received from node 5 when c was H in the prior period.

clock high When clock c goes H the *master* latches, and the *slave* becomes transparent. With TG$_1$ off and TG$_2$ on, the master positive feedback loop (nodes 4, 5, 6) holds the value received when c was L before going H. Since TG$_3$ is on, and TG$_4$ is off, the value at node 5 is passed to nodes 7, 8, and 9 to become the next state of q.

clock low When clock c goes L the *master* becomes transparent, and the *slave* latches to hold the next state of q until clock goes H again. This ends one clock period.

Figure 412 JK flip-flop *TG*

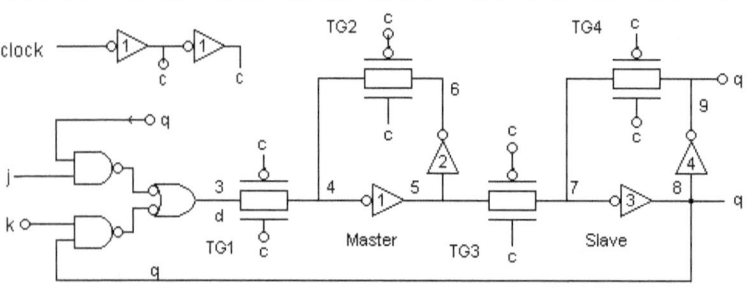

Design See D Flip-flop *TG*

5 Complex Digital Circuits with Memory

From here forward the emphasis is on circuits that are an assembly of identical cells, such as a 2 bit shift register cell. The integrated circuit layout of an assembly of identical cells is an orderly, repetitive pattern – *a systematic layout*. Systematic layouts are intrinsically free of layout errors.

The upcoming circuits are suitable for computer data paths and computer controllers (finite state machines) – circuits such as shift registers, storage registers under load control, storage registers on a bus, and programmable logic arrays (PLAs) as combinational logic.

5.1 Two Phase Clocks

Data is processed in all cells, in parallel, during each clock period. During processing the circuits in each cell are isolated from circuits in other cells. Isolation is achieved by the *clock waveforms*. One clocking scheme is referred to as a *two phase non overlapping clock* (Figures 501, 502). Two synchronized waveforms form the two phases. When one phase is high the other is low. The high state phases are separated by *both* waveform phases being in the low state. When a phase is high one or more transmission gates (TG's) are turned on. When both phases are low all TG's are off. Consequently circuits are isolated from adjacent cells (e.g. Figure 504a page 105). Add an inverter to the latch layout (Figure 402e page 88) to produce a two phase clock layout.

Figure 501 Two phase non overlapping clock

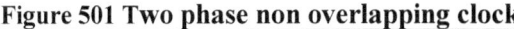
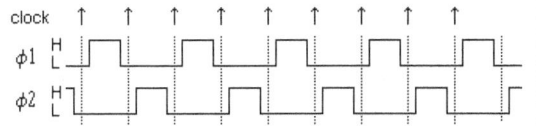

Figure 502 Two phase clock circuit

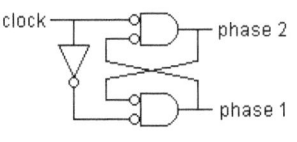

Problem 501 Draw a layout for the two phase clock.

5.2 Dynamic Shift Register

A *shift register* is a chain of one-bit storage circuits. The shift register has the property that all bits stored in the register can move, on the same clock edge, one position to the left (or right) of their present position in the chain by asserting control lines. A *register* is a group of n flip-flops, or equivalent, storing n bits that relate to each other in some way as defined by the wiring of control logic. A *serial shift right* register has the output of bit n connected to the input of bit n+1 (node 3 Figure 504a), and whose *next state defining equation* would be $d_{n+1}=q_n$ if we were using d flip-flops. The same equation applies when we use elementary dynamic logic as a substitute for d flip-flops (Dynamic Logic Circuits – the Basic Idea).

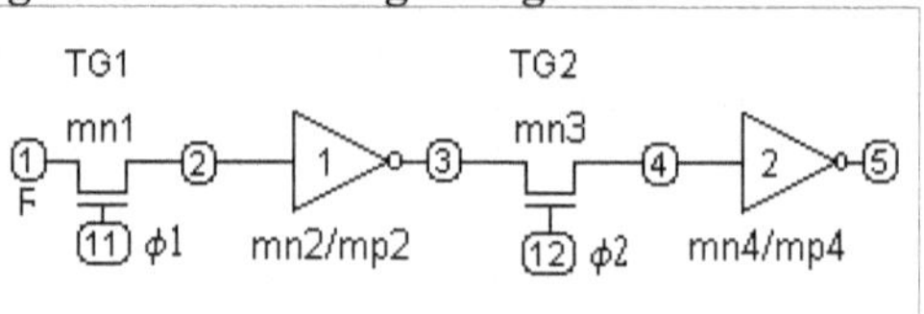

How it works Data is stored in the gate capacitor of each inverter, as well as the drain capacitor of a TG (nodes 2 & 4, Figure 504a). A bit is transferred from left to right each time ϕ_1 or ϕ_2 goes H. In what follows we say "nodes", because a shift register has n cells. Observe that node 5 is node 1 of the next cell.

Figure 504a Shift Right Register - 2 Bit Cell

With ϕ_1 at H and ϕ_2 at L, nodes 1 are connected to associated nodes 2 in all cells (Figure 504a). Data moves from nodes 1 to inverter output nodes 3 in every cell.

With ϕ_1 at L and ϕ_2 at H, nodes 3 are connected associated nodes 4. Data moves from nodes 3 to inverter output nodes 5.

With ϕ_1 at L and ϕ_2 at L, all transmission gates are off in all cells so that nothing changes.

The shortest period (highest clock frequency) is determined by the time required to charge and discharge the capacitors at nodes 2 and 4 and inverter delay.

Emphasis – the use of inverters requires 2 one bit cells so that data out H equals data in H.

Dynamic Logic Circuits - The Basic Idea

The basic idea exploits the fact mos circuit nodes have no DC paths to any other node, or to ground. And, each node has some equivalent capacitance C to ground. Nevertheless there are parasitic leakage paths for charge.

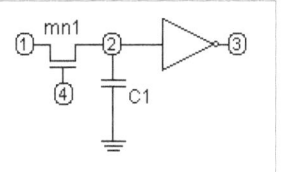

In a digital system node voltages are at L or H, where L=0 volts and H=V_{DD}. When V_4 goes H the mn_1 nmos transmission gate turns on and transfers V_1 to V_2. If V_1=H, the H at node$_1$ charges C_1 to $V_{DD}-V_{TN}$ at node$_2$. Or transfers L to L. Inverter output V_3 has the correct value, 0V or V_{DD}, in either case. When V_4 goes L the TG nmos transistor mn_1 turns off and node$_2$ holds (remembers) the last logic level be it L or H (0V or $V_{DD}-V_{TN}$). This not quite correct because there is a parasitic charge leakage path through mn_1 to the substrate. The following estimate shows that voltage on node$_2$ falls very, very slowly.

$$JS = 5 \cdot 10^{-9} \, A / m^2 \qquad AD = 5.5W \, \lambda^2$$

$$I_{leakage} = JS \times AD = 5 \cdot 10^{-9} \times 5.5W \, \lambda^2$$

$$If \; W = 10\lambda, \quad \lambda = 0.09 \, \mu m, \; then$$

$$I_{leakage} = 5 \cdot 10^{-9} \frac{A}{m^2} \cdot 10^{-12} \frac{m^2}{\mu m^2} \times 5.5 \cdot 10 \cdot (0.09)^2 \mu m^2 = 2.23 \cdot 10^{-21} \, A$$

$$\frac{dV}{dt} = \frac{I}{C} = \frac{2.23 \cdot 10^{-21}}{20 \cdot 10^{-15}} = 0.11 \frac{\mu V}{sec} = \frac{1V}{105 \; days}$$

A systematic layout There are 2 bits in each cell or u cell. The row and column layout repeats every two bits in the horizontal and vertical directions. A 32 bit word shift register has 32 horizontal shift register rows. If each row has 50 bits, then the shift register layout is an array 25 cells wide and 32 cells high (Figure 503). The u cell is an upside-down version of cell.

Figure 503 Shift Register Floorplan

	0	1	2		23	24
0	cell	cell	cell		cell	cell
1	u cell	u cell	u cell		u cell	u cell
2	cell	cell	cell		cell	cell
3	u cell	u cell	u cell		u cell	u cell
26	cell	cell	cell		cell	cell
27	u cell	u cell	u cell		u cell	u cell
28	cell	cell	cell		cell	cell
29	u cell	u cell	u cell		u cell	u cell
30	cell	cell	cell		cell	cell
31	u cell	u cell	u cell		u cell	u cell

Spice Program 5041

```
Fig5041.ckt Shift Register Transient Response
.include 180_N1P1.txt

.subckt mn1 103 102 101 99
MN1  103 102 101 99 N1 L=0.18u  W=0.45u      ; 5/2=W/L
+ AD=0.223p AS=0.223p PD=1.89u PS=1.89u
.ends mn1

.subckt mp1 209 208 207 298
MP3  209 208 207 298 P1 L=0.18u  W=0.45u     ; 5/2=W/L
+ AD=0.223p AS=0.223p PD=1.89u PS=1.89u
.ends mp1

Vdd 98  0  DC 1.8
V11 11 0 PULSE(0 1.75 0n 100p 100p  0.4n 2n)
V12 12 0 PULSE(0 1.75 1n 100p 100p  0.4n 2n)
*.ic V(2)=0  V(4)=0
V1   1 0 PULSE(0 1.95 0p 100p 100p  0.6n 4n)

xmn1  2 11    1     0 mn1 ; 5/2
xmp2  3  2   98    98 mp1 ; 5/2
xmn2  3  2    0     0 mn1
xmn3  4 12    3     0 mn1
xmp4  5  4   98    98 mp1
xmn4  5  4    0     0 mn1

*.PLOT TRAN V(11) V(12) V(1)        0,2 :Figure 50411
*.PLOT TRAN V(11) V(12) V(2) V(3)   0,2 :Figure 50412
*.PLOT TRAN V(11) V(12) V(3) V(4)   0,2 :Figure 50413

.TRAN 1e-011 5e-009 0 1e-011
.TEMP 27
.PLOT TRAN V(11) V(12) V(3) V(4) 0,2
.PRINT TRAN V(4)
.end
```

Design Design is an iterative process, because adding transistors adds capacitors. The number of iterations can be minimized if we estimate this additional capacitance or we use a trick.

A useful time saving trick with arrays of cells is to start with minimum transistor sizes such as $W_p/W_n=5\lambda/5\lambda$. Then we observe the performance, and move on from there (Spice program Fig5041.ckt).

Figure 504a Shift Right Register - 2 Bit Cell

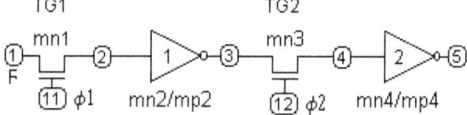

Input V_1 is sampled by V_{11} (ϕ_1) so that node 2 follows node 1 when ϕ_1 turns on TG1 (Figures 50411, 504a).

Figure 50411 Shift Register Input V_1, V_{11} phase ϕ_1, V_{12} phase ϕ_2

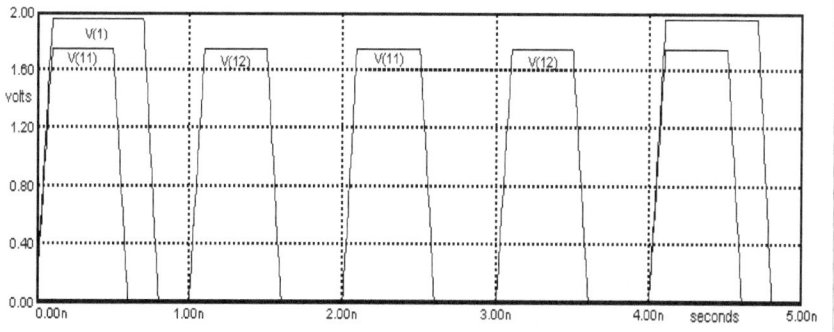

Figure 50412 Shift Register, node 3 is node 2 inverted

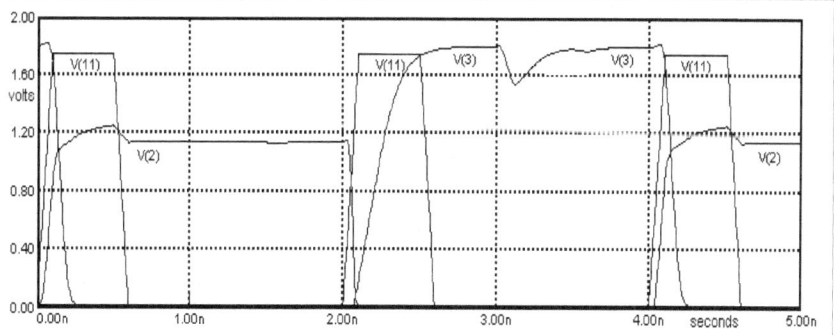

Time 0ns　Input V_1 goes to H and ϕ_1 (V_{11}) turns on TG_1. V_2 follows V_1 until it rises to $V_{DD}-V_{TN}$ when mn_1 of TG_1 turns itself off. V_2 holds at $V_{DD}-V_{TN}$ (Figure 50412).

Time 2ns　ϕ_1 (V_{11}) turns on TG_1, and V_2 follows V_1 by falling to L. V_2 holds at L. V_3 goes to H as the inverse of V_2 (Figure 50412, 504a).

Figure 50413 Shift Register, node 3 transfer to node 4 by phase 2

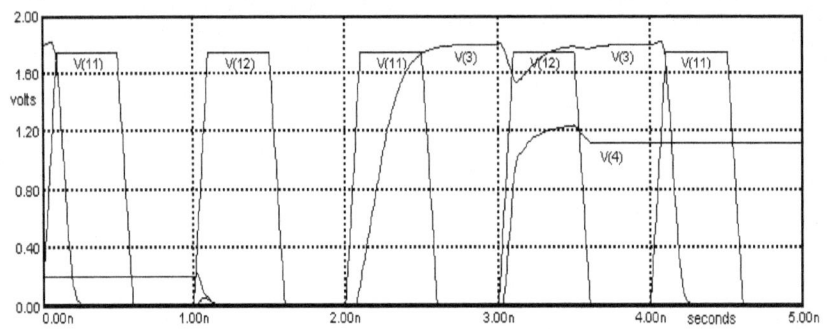

Time 3ns Input V_3 to TG_2 is sampled by V_{12} (ϕ_2) so that node 4 follows node 3 when ϕ_2 turns on TG_2 (Figures 50413, 504a).

Shift Register Cell Layout A layout plan starts with a transistor count. The Shift Register cell has 2 pmos and 4 nmos transistors. The 4 nmos transistors are placed in a p-well (anywhere on the p-wafer). The 2 pmos transistors are placed in n-wells (that you create).

The pmos transistors are placed in a row whose area is defined as 1 n-well. The nmos transistors are placed in a row below the pmos transistors. (Figure 504b). This placement is consistent with the pmos-nmos CMOS circuit structure.

Figure 504b Shift Register Cell

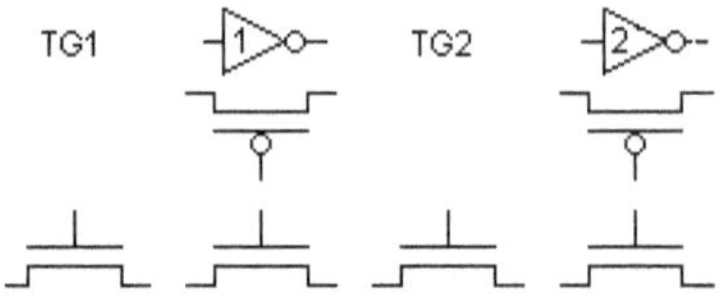

TG1   TG2

Active boxes define a transistor's source and drain areas. Metal$_1$ to poly connections are made by vias that are plated holes in the oxide separating the poly and metal$_1$ layers. Small boxes represent the vias connecting poly to metal$_1$, & active to metal$_1$.

Figure 504c Shift Register Cell Layout

TG1   TG2

poly   mp2   mp4

1   2   3   4   5

mn1   mn2   mn3   metal1

mn4

ph1   ph2

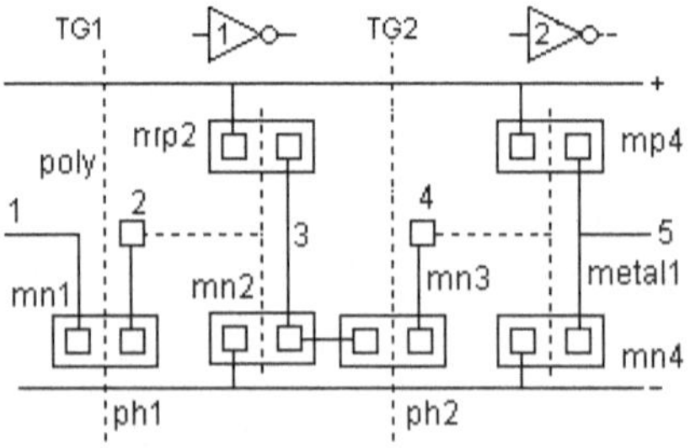

Dotted poly and solid metal$_1$ lines defining the inverters' pmos-nmos pairs are added (Figure 504c). Then poly and metal$_1$ lines are added as specified by the rest of the Shift Register circuit schematic (Figure 504a).

In this layout V_{dd} and 0 volts are horizontal metal$_1$ lines. Poly over active defines the gate of an mos transistor (Section 1.3.1 page 11). V_{dd} connects to the source of inverter pmos mp$_2$. Drain of mp$_2$ connects to mn$_2$'s drain, whose source is connected to 0 volts. Inverter mp$_4$, mn$_4$ is wired in the same way. Nodes 2 and 4 are inverter inputs, and nodes 3 and 5 are inverter outputs. Note the 5 wire that is intended to connect to the node 1 contact of the next cell. Figure 504c shows mn$_1$ and mn$_3$ TGs turned on and off by phase ϕ_1 and phase ϕ_2 poly lines. However metal$_2$ lines are used in the layout (Figure 504d).

An upside-down u cell has the 0 volts line at the top. Clearly a cell and a u-cell below it can be merged by merging the 0 volt lines (or just placing them side by side). Then the cell/u-cell pair is copied and merged with the cell/u-cell pair above it by merging the V_{dd} lines, and so forth until the 1 cell wide column array has 32 rows. Then the 1 wide, 32 high array is copied to create a 25 wide, 32 high array. Correctly positioned copies do *not* produce layout errors. Note that in a cell/u-cell column the vertical phase ϕ_1 and ϕ_2 metal$_2$ lines connect the gates of all mn$_1$, mn$_3$ TG transistors (Figure 504d).

Figure 504d Shift Register Cell Layout 31λ x 67λ (dots 2λ)

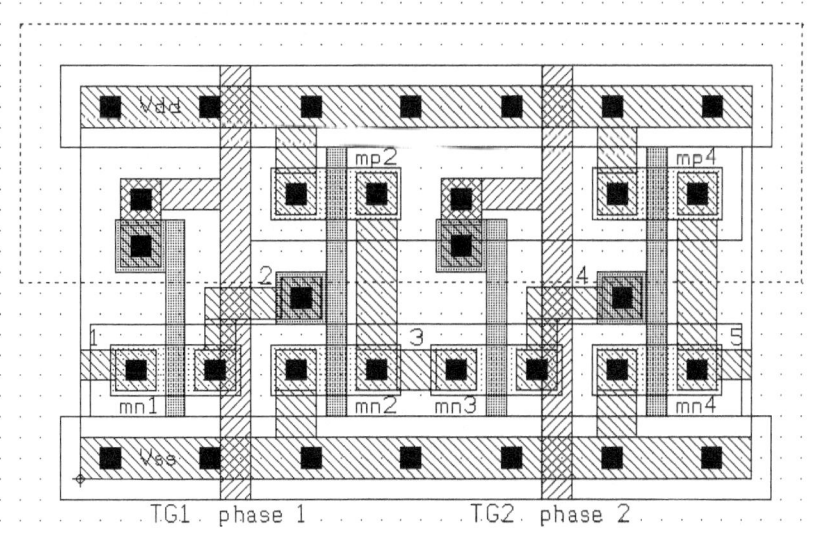

Problem 502 Reference Figure 504d. Draw the Lasi layout.
Problem 503 Reference Figures 503, 504d. Draw a rank 3 shift register layout for a 3 row by 4 column array of cells.

5.3 Dynamic Storage Register

A register is a memory storing n bits where each bit is stored in a one bit storage cell (Figure 506a). Data is written into the register's storage cells by asserting *load* command g (Figure 505). Data in the register changes on a clock edge if load input g is asserted. The defining equation

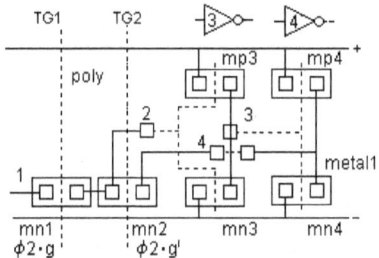

Figure 506a One bit storage cell

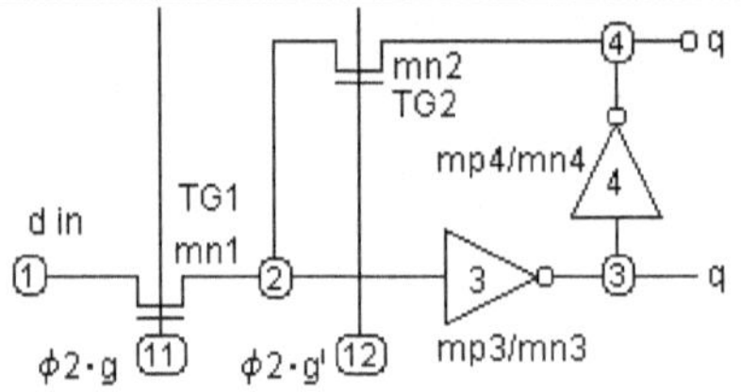

is $q^+=gd+g'q$ (load data d, or load present state q). The one bit storage cell retains its state for an indefinite number of clock cycles until reloaded by another load g. Observe that the ϕ_1 and ϕ_2 control lines are perpendicular to the bit cell rows for systematic layout capability.

Figure 506b One bit storage cell 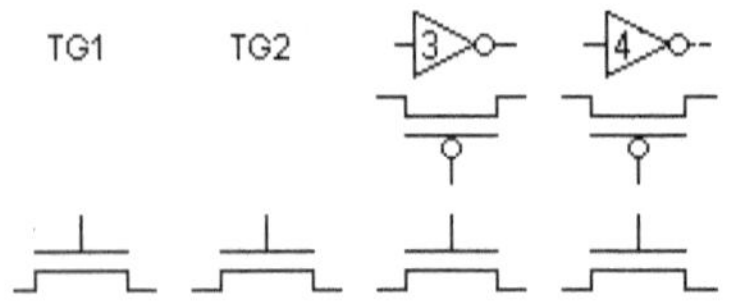   **Figure 506c One bit storage cell**

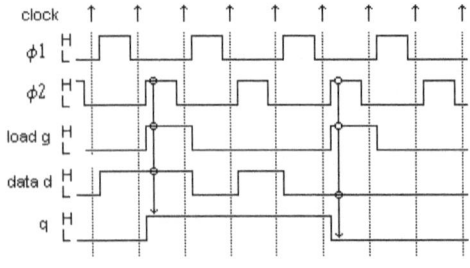

When ϕ_2 and g go H, mn_1 turns on, and data d is transferred to nodes 2, 3, 4 (Figures 505, 506a). Mn_2 is off, because $\phi_2 g'=0$ (L). Stored data is refreshed in each cycle when $\phi_2 g'=1$ (H). Here are the defining equation and truth table with table entered variable q.

Figure 505 Load command action

$$q^+ = gd + g'q$$

g	d	q^+	action
0	0	q	store q
0	1	q	"
1	0	0	store d
1	1	1	"

Problem 504 Draw a layout for the one bit storage cell (Figure 506d).
Problem 505 Draw a layout for a rank 3 4 row by 4 column array of cells.

In a cell/u-cell column the vertical phase 1 (ϕ_2 g) and phase 2 (ϕ_2 g') metal$_2$ lines connect the gates of all mn$_1$, mn$_2$ TG transistors (Figure 506d).

Figure 506d Register with Load 37λ high (dots 1λ)

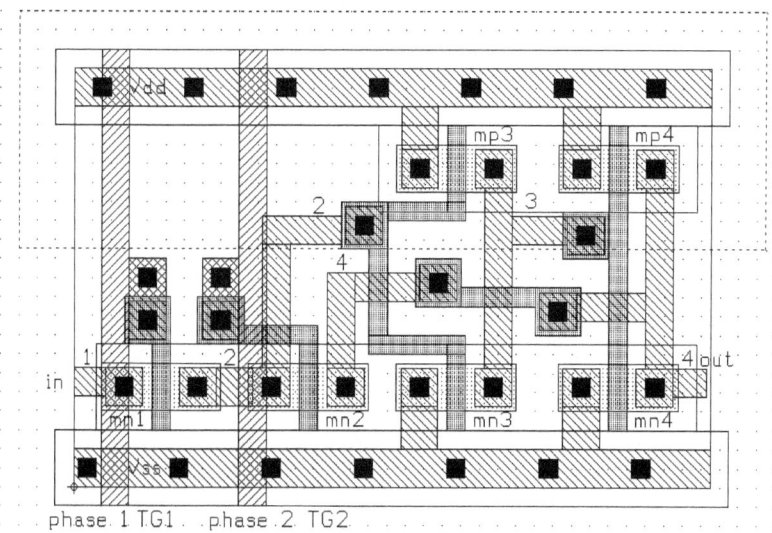

Figure 50611 Storage Register Input V$_1$, V$_{11}$ phase ϕ_1, V$_{12}$ phase ϕ_2

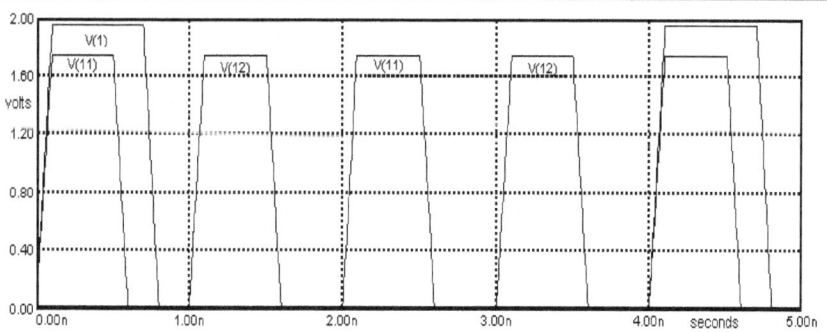

Figure 50612 Storage Register, node 2 transfer to node 3 by phase 2 V$_{12}$

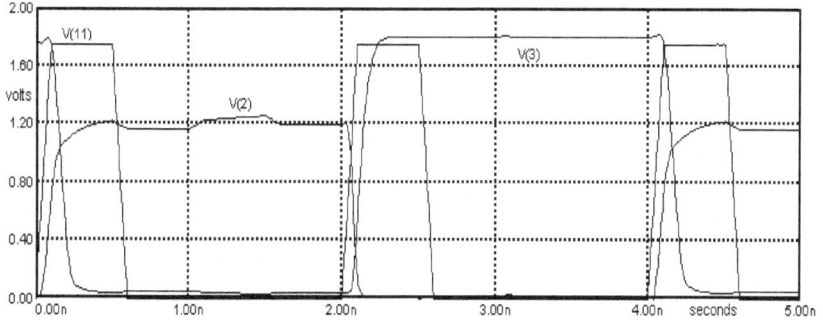

Figure 50613 Storage Register, inverted node 3 to node 4

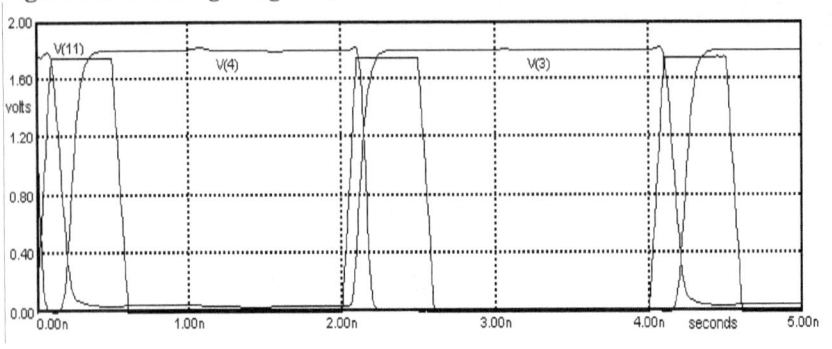

Spice Program 5061
Fig5061.ckt Storage Register Transient Response
.include 180_N1P1.txt

```
.subckt mn1 103 102 101 99
MN1  103 102 101 99 N1  L=0.18u  W=0.45u     ; 5/2=W/L
+ AD=0.223p AS=0.223p PD=1.89u PS=1.89u
.ends mn1

.subckt mp1 209 208 207 298
MP1  209 208 207 298 P1  L=0.18u  W=0.45u    ; 5/2=W/L
+ AD=0.223p AS=0.223p PD=1.89u PS=1.89u
.ends mp1

Vdd 98  0  DC 1.8
V11 11 0 PULSE(0 1.75 0n 100p 100p  0.4n 2n)
V12 12 0 PULSE(0 1.75 1n 100p 100p  0.4n 2n)
.ic V(2)=0  V(4)=1.8
V1   1 0 PULSE(0 1.95 0p 100p 100p  0.6n 4n)

xmn1 2 11   1   0 mn1
xmn2 4 12   2   0 mn1
xmp3 3  2 98 98 mp1
xmn3 3  2  0   0 mn1
xmp4 4  3 98 98 mp1
xmn4 4  3  0   0 mn1

*.PLOT TRAN V(11) V(12) V(1) 0,2        ; Fig50611
*.PLOT TRAN V(11) V(12) V(2) V(3) 0,2   ; Fig50612
*.PLOT TRAN V(11) V(12) V(3) V(4) 0,2   ; Fig50613
.TRAN 1e-011 5e-009 0 1e-011
.TEMP 27
.PLOT TRAN V(11) V(12) V(3) V(4) 0,2
.end
```

5.4 Registers on a Bus

Microcode command *ldz* from the controlling cpu stores a data word that is on the z bus in a 32 bit register. The cpu can read out the word onto the x or y busses by microcode commands *selx* and *sely* (Figure 507).

Example Register loads and tri-state outputs are controlled by ϕ_1 and ϕ_2 (Figure 507). The present values are loaded by $\phi_1 \times ldz$, and latched by ϕ_2.

Figure 507 Register - tri-state output, xyz busses

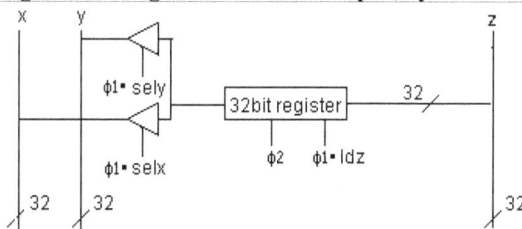

Each bit of any word is stored in a one bit storage cell (Figure 508a). Data is written into the register's storage cells by asserting a *ldz* input. The *ldz* signal turns on transmission gate mn_1 that connects bit line z_k to cell input node 2. The bit stored in the cell is refreshed by ϕ_2. The one bit storage cell retains its state for an

Figure 508a 1 bit register tri-state outputs

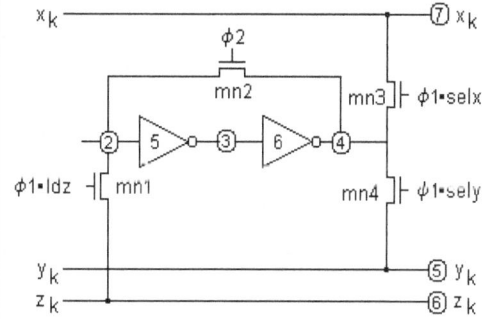

indefinite number of clock cycles until reloaded by another load signal. Data is read from a cell by *selx and sely* commands that connect bit outputs to the x and y busses via the mn_3 and mn_4 TGs.

One-bit Register Layout The layout is started by positioning transistors (Figure 508b). Then the two inverters are drawn with horizontal metal V_{dd} and ground lines, and three horizontal metal bus lines x, y, z are added (Figure 508c). The *ldz* TG mn1 is added at the left, and connected to the z bus at node 6 by a poly line that crosses over the z, y, and 0V metal lines. The ϕ_2 *feedback TG mn2*, and the *selx* and *sely* TGs are added at the right. Metal$_1$ connects node 4 to the inputs to the three TGs mn2, mn3, mn4.

The *feedback* line from the mn2 output connects to node 2. The mn3, mn4 outputs connect to *bus x* and *bus y* by metal to poly vias, and poly lines that then cross metal lines. Transistor gates are vertical to be compatible with vertical metal$_2$

Figure 508b One bit storage cell

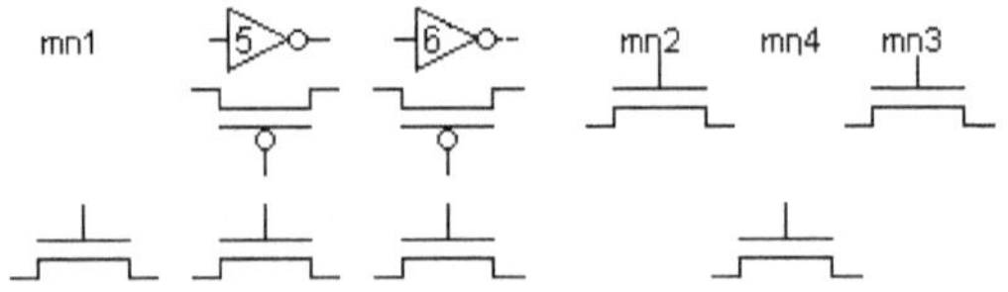

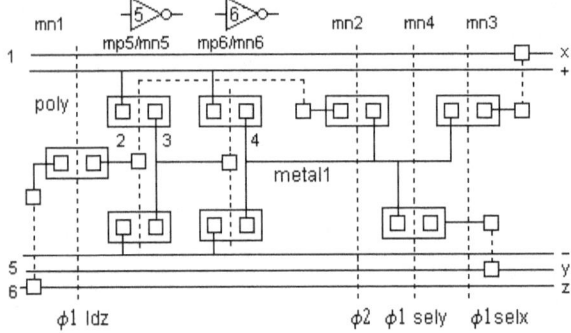

control lines. Control lines use ϕ_1 and ϕ_2 (Figure 508c shows poly control lines). Metal$_2$ control lines are used in the layout.

Figure 508c One bit storage cell

A register cell's metal$_2$ control lines extent 1.5λ beyond the bus lines. Then busses in adjacent cells are separated by 3λ according to design rule 5.2.

Problem 506 Draw a layout for the one bit storage cell (Figure 508d).
Problem 507 Draw a layout for a rank 3, 4 row by 4 column array of cells.

Figure 508d One bit storage cell, tri-state output, xyz busses (dots 2λ)

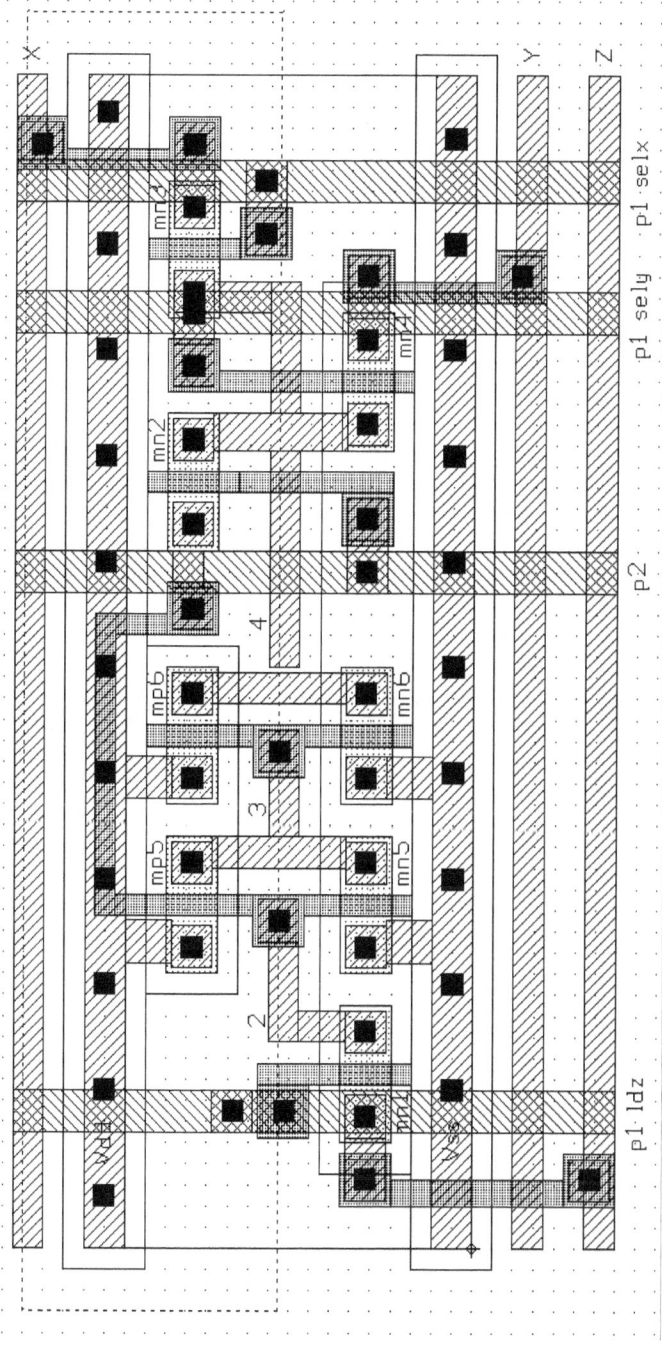

Spice Program 5081

Fig5081.ckt Storage Register Transient Response

.include 180_N1P1.txt

```
.subckt mn1 103 102 101 99
MN1  103 102 101 99 N1  L=0.18u  W=0.45u       ; 5/2=W/L
+ AD=0.223p AS=0.223p PD=1.89u PS=1.89u
.ends mn1

.subckt mp1 209 208 207 298
MP1  209 208 207 298 P1  L=0.18u  W=0.45u   ; 5/2=W/L
+ AD=0.223p AS=0.223p PD=1.89u PS=1.89u
.ends mp1

Vdd 98  0  DC 1.8

V11 11 0 PULSE(0 1.75 0n 100p 100p  0.4n 2n)
V12 12 0 PULSE(0 1.75 1n 100p 100p  0.4n 2n)

.ic V(2)=0  V(4)=1.8
V1   1 0 PULSE(0 1.95 0p 100p 100p  0.6n 4n)

xmn1 2 11 1 0  mn1
xmn2 4 12 2 0  mn1

xmp5 3 2 98 98 mp1
xmn5 3 2  0  0 mn1

xmp6 4 3 98 98 mp1
xmn6 4 3  0  0 mn1

xmn3 7 11 4  0 mn1
xmn4 5 12 4  0 mn1
C7 7 0 20f
C5 5 0 20f

*load inverter
xmp16 26 5 98 98 mp1
xmn16 26 5  0  0 mn1

*.PLOT TRAN V(2) V(3) V(4) 0,2
.TRAN 1e-011 5e-009 0 1e-011
.TEMP 27
.PLOT TRAN V(11) V(12) V(4) V(7) V(5) 0,2
.PRINT TRAN V(5)
.end
```

5.5 Programmable Logic Array

A logic function equation can be written as a sum of minterms (OR of ANDS), or as a product of maxterms (AND of ORs). One way to convert a logic function into a systematic layout structure is to use a memory to produce the function's complete truth table. The m logic function inputs are the memory's address lines that select n bit words to read one at a time. In many cases most of the memory chip area is wasted, because most of the 2^m addresses never occur when the logic function executes.

An alternative is a Programmable Logic Array (PLA) that implements an OR of ANDS. The PLA sum of minterms representation of a logic function has one AND circuit for each minterm, and one OR circuit for each output. In other words the PLA implementation has what is referred to as the AND array, and the OR array. The PLA has one input for each variable, and one output for each function. There is a one to one correspondence with the associated truth table inputs and outputs.

Here is how four elementary operators AND, OR, NOT, and XOR are implemented by a PLA (Figure 509 series).

Each OR circuit (1 to 4 inputs) is assembled from one pmos and one to three nmos (Figure 509b).

Truth Tables

Row	x	y	*AND* xy	*OR* $x+y$	*NOT* y'	*XOR* $x \oplus y$
0	0	0	0	0	1	0
1	0	1	0	1	0	1
2	1	0	0	1	1	1

Figure 509a PLA 2-input AND

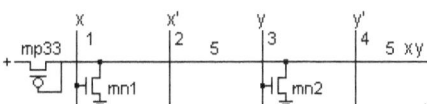

Figure 509b PLA OR

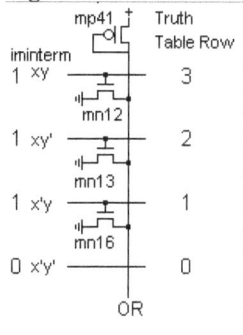

In the AND circuit (Figure 509a), when x AND y are L mn$_1$ and mn$_2$ are off, and mp$_{33}$ charges output line xy to H=V_{DD}−V_{TH}. If mn$_1$ and, or mn$_2$ are on the output line is at L=V_{SS}. In a 0/1.8volt system the output voltages are H=1.3V and L=0V.

Spice Program 5111 PLA AND array

```
Fig5111.ckt PLA AND array Register Transient Response
* only minimum size transistors are used

.include 180_N1P1.txt
.subckt mn1 103 102 101 99

MN1 103 102 101 99 N1 L=0.18u W=0.45u      ; 5/2=W/L
+ AD=0.223p AS=0.223p PD=1.89u PS=1.89u
.ends mn1
.subckt mp1 209 208 207 298

MP1 209 208 207 298 P1 L=0.18u W=0.45u      ; 5/2=W/L
+ AD=0.223p AS=0.223p PD=1.89u PS=1.89u
.ends mp1

Vdd 98 0 DC 1.8

V23 23 0 PULSE(1.85 0.10 000p 100p 100p 2.4n 5n)
V24 24 0 PULSE(0.15 1.75 000p 100p 100p 1.15n 2.5n)

*x input
xmn23 1 98 23 0 mn1    ; 5/2
xmp21 2 1 98 98 mp1    ;12/2
xmn21 2 1 0 0 mn1

*y input
xmn24 3 98 24 0 mn1
xmp22 4 3 98 98 mp1
xmn22 4 3 0 0 mn1
*matrix
xmn1 5 1 0 0 mn1
xmn2 5 3 0 0 mn1
xmn3 6 1 0 0 mn1
xmn4 6 4 0 0 mn1
xmn5 7 2 0 0 mn1
xmn6 7 3 0 0 mn1
xmn7 8 2 0 0 mn1
xmn8 8 4 0 0 mn1
*AND pullups
xmp33 98 5 5 98 mp1
xmp32 98 6 6 98 mp1
xmp31 98 7 7 98 mp1
xmp30 98 8 8 98 mp1
*.ic V(5)=1.8 V(6)=1.8

C6 6 0 2.8f
```

*OR pullups
xmp40 98 11 11 98 mp1
xmp41 98 12 12 98 mp1
xmp42 98 13 13 98 mp1
xmp43 98 14 14 98 mp1

*OR matrix
xmn11 11 5 0 0 mn1
xmn12 12 5 0 0 mn1
xmn13 12 6 0 0 mn1
xmn14 13 6 0 0 mn1
xmn15 14 6 0 0 mn1
xmn16 12 7 0 0 mn1
xmn17 14 7 0 0 mn1
xmn18 13 8 0 0 mn1

*output inverters omitted

*.PLOT TRAN V(5) V(6) V(7) V(8) 0,2
*.PLOT TRAN V(1) V(4) 0,2
.TRAN 1e-011 5e-009 0 1e-011
.TEMP 27
.PLOT TRAN V(23) V(24) 0,2
.PRINT TRAN V(6)
.end

Figure 511 PLA MOS Circuit

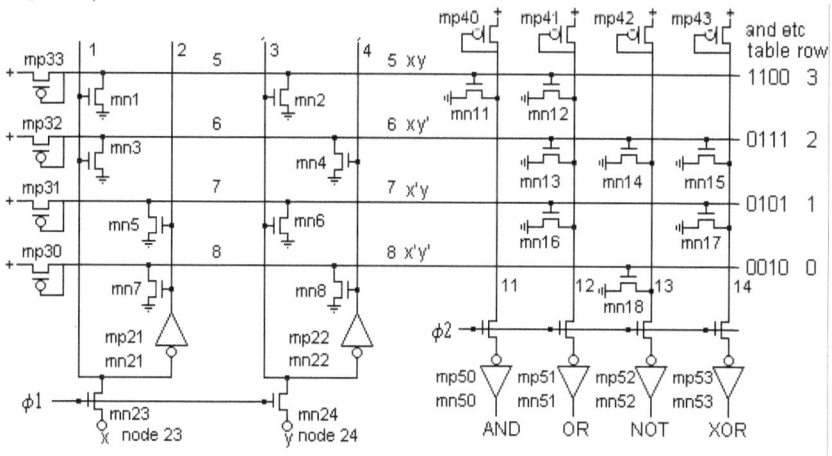

In the OR circuit (Figure 509b), when all minterm inputs are L the mn are off, and mp_{41} charges output line H=1.3V. When any input is H the mn is on and the output is L or 0V.

Figure 51111 PLA x, y input signals node 23 V_{23} and node 24 V_{24}

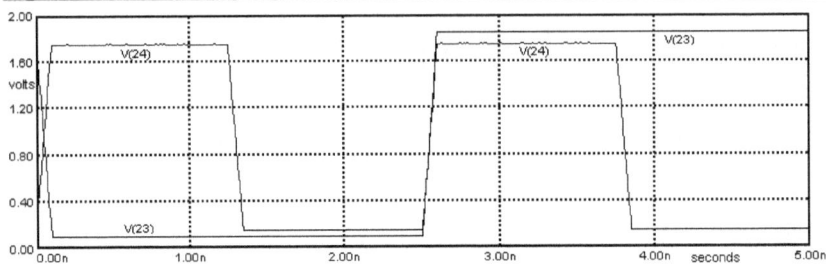

Figure 51112 PLA x and y' signals node 1 V_1 and node 4 V_4

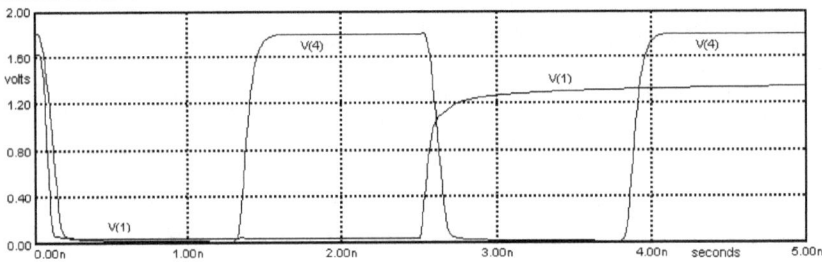

Figure 51113 PLA AND output nodes 5, 6, 7, 8 when xy, xy', x'y, x'y' are true

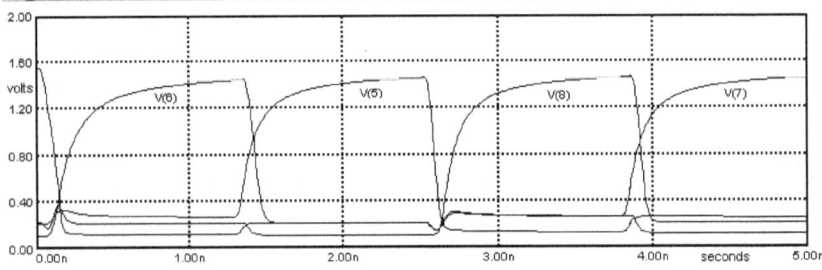

The 2-input AND array has 4 rows. *A row implements one of the four minterms of two variables.* The input side of the PLA has active low x, y inputs to the ANDs (Figure 510a). Inverters make internal x', y' lines active low. Anticipating the chip layout, the x, x', y, and y' are vertical lines, and the AND gate output lines are drawn as horizontal lines

Figure 510a PLA without memory

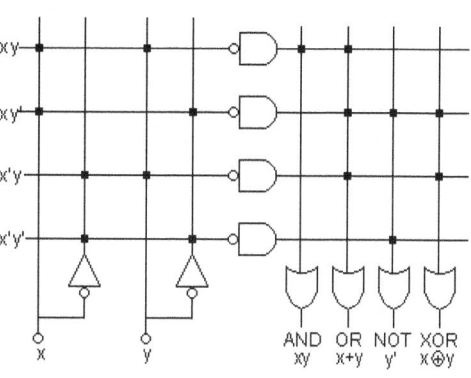

crossed by the x, x', y, and y' vertical lines. Transistors are represented by dots that connect vertical lines to horizontal lines.

Good practice dictates deriving a floorplan (Figure 510b) from the circuit (Figure 511). The idea is to decompose the circuit into a small number of basic circuit cells that are used many times. Decomposition is *not* unique. The original

Figure 510b PLA MOS Circuit Floorplan

+	0volts		5	5	5	5	+
2	4	4	7/6	7/7	6/7	6/7	
2	4	3					
2	3	4	6/6	7/6	6/7	7/6	
2	3	3					
φ1 →	1	1	8	8	8	8	← φ2
	x	y	AND	OR	NOT	XOR	

list of 10 cells (1 to 8) was supplemented by three transition cells (9, 10, 11) whose need became clear as the actual layout progressed.

cell	function	cell	function
1	input to ands	7/6	or array nmos/omit
2	and pull up	7/7	or array nmos/nmos
3	and array omit/nmos	8	output inverter
4	and array nmos/omit	9	metal to poly transition
5	or pull up	10	metal to metal transition
6/6	or array omit/omit	11	active to metal transition
6/7	or array omit/nmos		

AND array layout The AND array layout Figure 511g (page 123) of the floorplan (Figure 510b) is an assembly of cells 1, 2, 3, 4 (Fig 511 c, d, e, f).

The PLA basic structure has active low input variables x, y on vertical wires (Figure 511a x, y input bubbles). The inverter outputs 21, 22 are active low x', y'. The AND outputs on horizontal wires go to H when all AND input nmos are off (e.g. xy = LL and mn_1, mn_2 are off), because always-on pmos $mp_{33, 32, 31, 30}$ have gate connected to drain.

Figure 511a PLA AND array

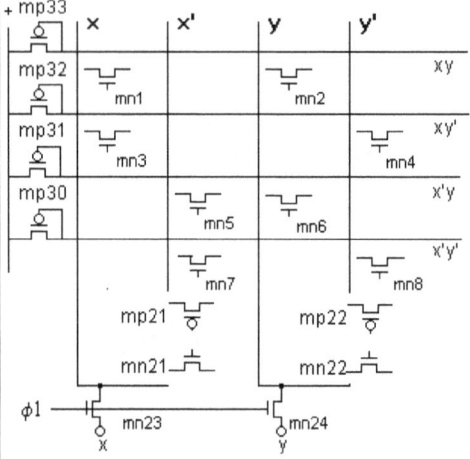

Extended nmos active boxes are used as wires connecting the nmos sources to ground.

Phase 1 TGs connect inputs x and y to the AND array.

x, y inputs are active low (lines 1, 3).

At inverter outputs x', y' inputs are active low (lines 2, 4).

Pmos are diode connected to act as pullups implementing the AND function (lines 5, 6, 7, 8).

Figure 511b PLA AND array

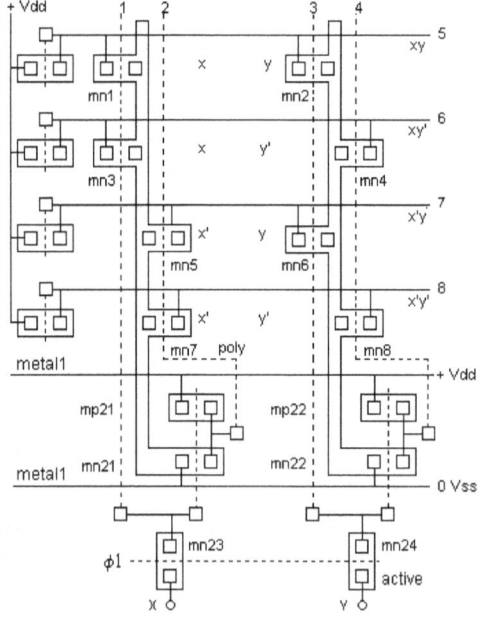

Draw pullup cell #2 (Figure 510b). The vertical column of four #2 cells (Figures 511b) requires a vertical metal V_{dd} line connected to the pmos sources on the left. The mos drains are connected to the poly gates by a metal$_1$-contact-poly transition pad (Figure 511c). The mos drain is also connected to the start of a horizontal metal output line that will connect to mos drains in cells #3, #4, and an OR input.

Figure 511c Cell #2 (1λ)

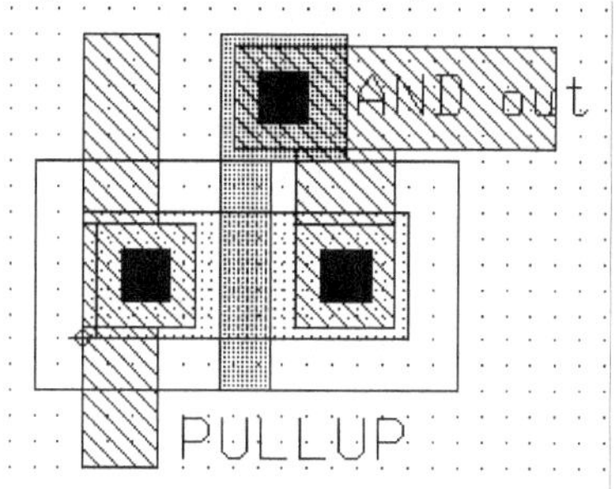

Observe (Figure 511b) that cells 3 and 4 are symmetrical (nmos is on the left or the right of the grounded active lines). So we draw cell #3 (Figure 511d) with one transistor at the right (Figure 511b mn$_4$), and cell #4 (Figure 511e) with one transistor at the left (Figure 511b mn$_3$). The height of cells #3 and #4 has to match the 17λ height of pullup cell #2 (Figure 511c). The cell widths are arbitrary at this point.

Cells 3 and 4 each have 2 *poly* lines as inputs x or x', y or y', 1 *active* ground line and 1 *metal$_l$* pullup line (AND out). The pullup line that starts in cell 2 is horizontal. The logical and ground lines have to be vertical in cells 3, and 4. This requires the vertical lines to be poly or active. This is why vertical input lines are poly to match the poly mos gates, and the vertical grounded active grounds the mos sources.

Observe that the mos transistors active to poly spacing is 1λ or more (rule 3.5 page 39).

Figure 511d Cell #3 (dots 1λ)

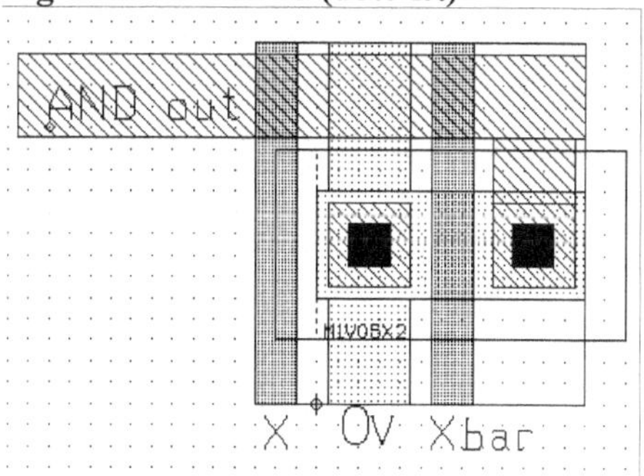

Figure 511e Cell #4 (dots 1λ)

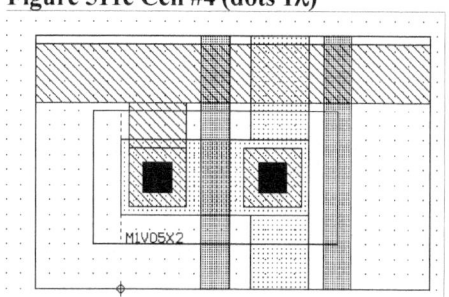

Start the AND array drawing by adding 4 #2 Cells at the left. Then place a #4 cell next to the mp_{32} #2 cell (Figure 511a). Place a #3 cell below the #4 cell. Align the poly lines. Extend the #3 cell to the left to touch the #2 cell. Select right hand boundaries so that #3 and #4 cells have the same width, which will need to be adjusted.

The floorplan shows that cell #3 and #4 widths have to match the width of inverter cell #1.

By now you might agree that drawing cells is an art form.

Start the input cell #1 drawing (Figure 511f) by drawing the inverter horizontal V_{dd} and V_{ss} metal$_1$ lines above and below pmos and nmos transistors. The pmos source is connected to V_{dd}, the pmos drain is connected to the nmos drain, the nmos source is connected to 0 volts. A metal$_1$ to poly transition pad connects the inverter output to a vertical output poly line z' driving nmos gates (lines 2 and 4 Figure 511b). A vertical poly line z drives nmos gates (lines 1 and 3). A 0 volts vertical active line grounds the AND array nmos sources. Cell #1 is 29.5λ wide. Make cell #3 and cell #4 29.5λ wide

Figure 511f Cell #1 (dots 1λ)

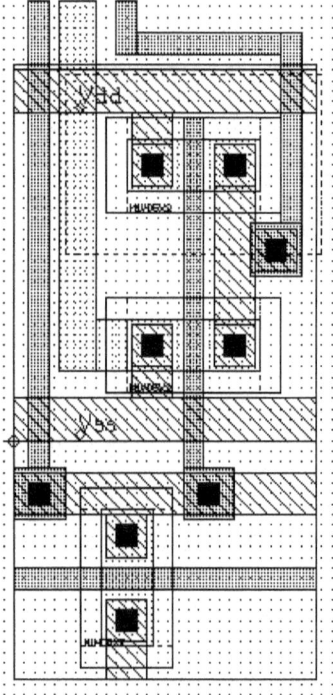

An nmos connected as a TG controlled by $\phi1$ is connected to the inverter gate poly line, and the active L z poly line (z is actually x or y).

PLA AND Array Layout
Add the column of cells #4, #4, #3, #3 next to the #2 cells (Figure 510b). Add a #1 cell below the column of 4. Align the poly columns.

Add the column of cells #4, #3, #4, #3 to the right (Figure 510b). Add a #1 cell below the column of 4. Align the poly columns.

The process is not easy. Many adjustments have to be made so that all cells line up as they should (Figure 511g).

Figure 511g PLA AND Array Layout (dots 2λ)

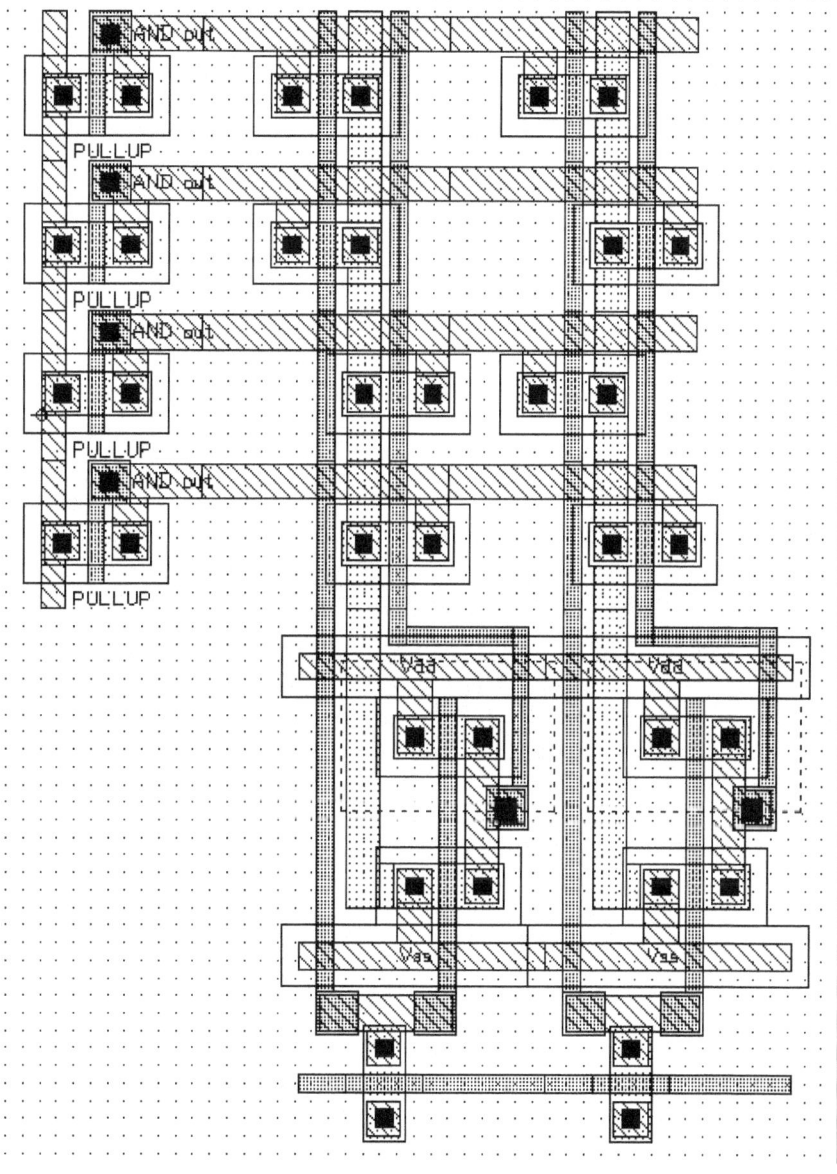

Problem 508 Draw layouts for cells #2, #3, #4 (Figures 511c, d, e).
Problem 509 Draw a layout cell #1 (Figure 511f).
Problem 510 Draw a layout of the AND array (Figure 511g).
Problem 511 Draw a layout of the OR array.

6 Analog Circuits

A current mirror, differential amplifier, operational amplifier, resistors, and capacitors are designed and their performance is evaluated by Spice. The circuits are based on two widely used analog circuits: the current mirror, and differential amplifier. They are the basic building blocks for analog integrated circuits. The performance of all circuits is evaluated by Spice.

In linear circuits MOS transistors are used in the constant current saturation mode (equation 2 long). There are long and short channel MOS transistors as explained in Mobility Short Channel (page 50). For example in a 0.18 μm process a transistor is short when channel length $L=L_{min}= 2\lambda = 0.18$ μm and voltage=1.8V. In the same process a transistor is long when channel length $L=10\lambda=0.90$ μm and voltage=1.8V. An MOS transistor is long when the values of L and supply voltage produce an electric field in the channel less than the critical value E_c (page 50).

> *A designer uses short L for large bandwidth & for high speed, wide W for high current, and long L when higher output resistance is required.*

MOS analysis requires a value for $\mu_n C_{ox}$, which appears in the I_{DS} equations. We start with μ_n=230 and μ_p=80 cm²/Vs mobility μ numbers for t_{oxide}=4nm, $V_{GS} \approx 0.5V$ where $L_{MIN}=0.18\mu m=2\lambda$ (Figures M302, M303 in Mobility Short Channel page 50). *The mobility μ is a function of V_{GS}.* As the design progresses we change the values 230 and 80 *as V_{GS} changes.*

$$(1a) \quad \mu_n C_{ox} = 230 \frac{cm^2}{Vs} \cdot 10^8 \frac{\mu m^2}{cm^2} \cdot 8.63 \cdot 10^{-15} \frac{F}{\mu m^2} = 200 \frac{\mu A}{V^2}$$

$$(1b) \quad \mu_p C_{ox} = 80 \frac{cm^2}{Vs} \cdot 10^8 \frac{\mu m^2}{cm^2} \cdot 8.63 \cdot 10^{-15} \frac{F}{\mu m^2} = 69 \frac{\mu A}{V^2}$$

The values of λ, W and L define the area of an MOS transistor. The area is $W(L+11\lambda)+4\lambda L$ as specified by the design rules (Figure 203). In all circuits we seek a minimum area, because what a wafer costs to produce is independent of whatever circuits are on the wafer.

Figure 203 Transistor Layout

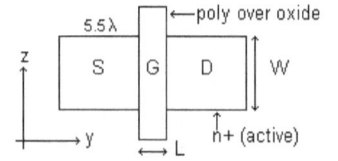

6.1 Current Mirror

Constant current sources have numerous applications in discrete and integrated MOS transistor circuits. The current mirror is a convenient way to produce constant current sources, while eliminating many resistors that require relatively large chip areas. Current mirror limitations are finite output impedance and operation over a range of voltages that is not rail to rail (V^+ to V^-). There are more complex mirror circuits that have fewer limitations.

How it works at DC The two transistor circuit mn_1, mn_2 as wired forms a current mirror (Figure 601). I_{DS1} is set by R_1 (see sidebar *Nmos replaces current mirror resistor page 130*). Diode connected nmos transistor mn_1's drain is connected to the gate that forces V_{DS} to equal V_{GS}. Transistor mn_2's gate is connected to mn_1's gate so that I_{DS_mn2} = constant × I_{DS_mn1} (equation 5). Equation 3 shows that if I_{DS} is selected, then V_{GS} is fixed.

Figure 601 Current Mirror

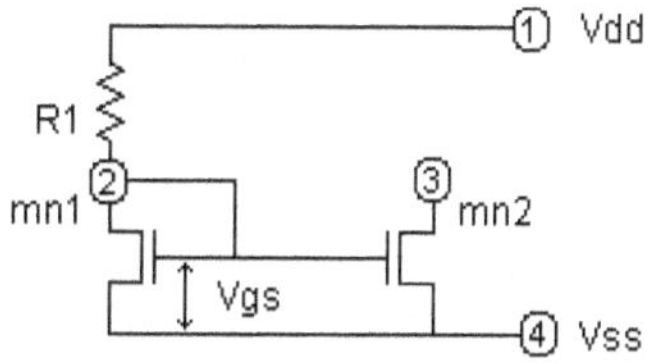

Saturation region, long channel

$$(2) \quad I_{DSn} = \frac{1}{2}\mu_n C_{ox}\frac{W}{L}(V_{GS}-V_{Tn})^2 = \frac{1}{2}\beta(V_{GS}-V_{Tn})^2$$

$$(3) \quad V_{GS} = V_{Tn} + \sqrt{\frac{2I_{DSn}}{\mu_n C_{ox}\frac{W}{L}}} = V_{Tn} + \sqrt{\frac{2I_{DSn}}{\beta}}$$

$$(4) \quad R_1 = \frac{V_{DD}-V_{SS}-V_{DS}}{I_{DS}} = \frac{V_{DD}-V_{SS}-V_{GS}}{I_{DS}}$$

The resistor is required, because the power supply voltage $V_{DD}-V_{SS}$ is a lot greater than $V_{DS1}=V_{GS1}$. The output current I_{DS2} is proportional to I_{DS1}.

$$(5) \quad I_{DS\_mn2} = \frac{1}{2}\mu_n C_{ox}\frac{W_2}{L_2}(V_{GS}-V_{Tn})^2 = \frac{W_2}{L_2}\frac{L_1}{W_1}\times\frac{1}{2}\mu_n C_{ox}\frac{W_1}{L_1}(V_{GS}-V_{Tn})^2$$

$$I_{DS\_mn2} = \frac{W_2}{L_2}\frac{L_1}{W_1}\times I_{DS\_mn1}$$

CMOS Circuit Design

Design: Voltages A current mirror has a voltage condition in addition to a current specification. A current mirror transistor inserted into a circuit has limits on the value of its V_{DS}. E.g. input nodes 2 and 6 are at 0 volts DC (Figure 605). The V_{GS} of mn_1 (V_2-V_3) plus V_{39} (V_3-V_9) the I_T mirror V_{DS}, add up to $|V_9|$. On the positive voltage side the mp_1 V_{GS} plus the mn_1 V_{DSsat} plus the positive output signal swing must be less than V_7.

Figure 605 Differential Amplifier

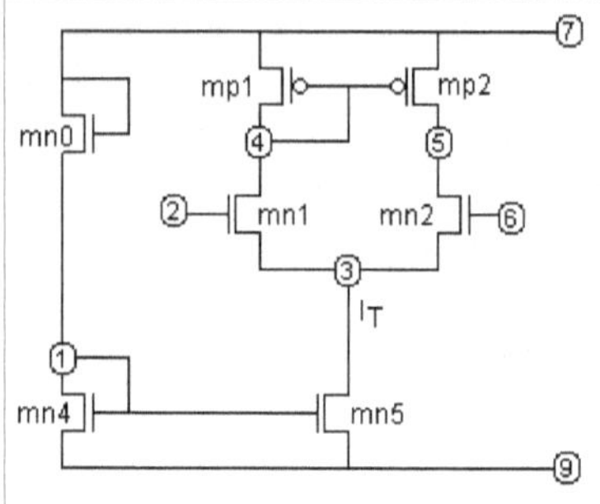

If ±1.8V power supplies are used, then we want all transistor's V_{GS} less than say 0.6=1.8/3 volts. Since V_{GS} decreases as W *increases* (equation 3) we usually use W/L>1 for the transistors.

Design: R_1 The value of the current mirror $I_{DS}=I_T$ is set by the circuit using the mirror. We want L=5L$_{min}$, because long L produces higher output resistance r_0 (Figure 60231 page 128). Parameter $\mu_n C_{ox}$ value is from equation 1a. Here is how W/L affects V_{GSn}.

Assume $I_{DS}=100\mu A$, W/L=10, L=5L$_{min}$=10λ=0.9μm.

$$(6a)\ V_{GSn} = V_{Tn} + \sqrt{\frac{2I_{DSn}L}{(\mu_n C_{ox})W}} = 0.45 + \sqrt{\frac{2\cdot100\cdot10^{-6}\cdot10\lambda}{(200\cdot10^{-6})\cdot100\lambda}} = 0.45+0.316 = 0.77V$$

$$(6b)\ R_1 = \frac{V_{DD}-V_{SS}-V_{GS}}{I_{DS1}} = \frac{1.8-(-1.8)-0.77}{0.1mA} = \frac{2.83}{0.1\cdot10^{-3}} = 28.3K \approx 30K$$

Change to W/L=2 to increase V_{GS} from 0.77V to 1.16V

$$(7a)\ V_{GSn} = V_{Tn} + \sqrt{\frac{2I_{DSn}L}{(\mu_n C_{ox})W}} = 0.45 + \sqrt{\frac{2\cdot100\cdot10^{-6}\cdot10\lambda}{(200\cdot10^{-6})\cdot20\lambda}} = 0.45+0.71 = 1.16V$$

$$(7b)\ R_1 = \frac{V_{DD}-V_{SS}-V_{GS}}{I_{DS1}} = \frac{1.8-(-1.8)-1.16}{0.1mA} = \frac{2.44}{0.1\cdot10^{-3}} = 24.4K \approx 24K$$

Note: 30K and 24K are standard 5% resistor values.

Problem 601 $r_0=1/\lambda I_{DS}$. Calculate r_0 and λ from the plots in Figure 60211.

Design: Output Resistance   Output resistance $r_0=1/\lambda I_{DS}$.

Figure 602 Current Mirror Test Circuit

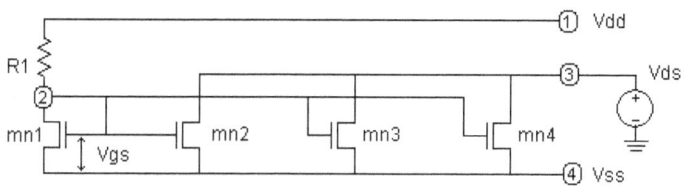

Select W/L=10 for mn$_1$ and W/L=5, 10, 20 for mn$_2$, mn$_3$, mn$_4$. to show the effect of W/L on r$_0$ Let $L=5L_{min}=10\lambda=0.9\mu m$ for all transistors (Figure 602). Increase V_3 from -1.8 to $+1.8$ volts and plot mn$_2$, mn$_3$, mn$_4$ I_{DS}. The plots (Figure 60211) show that as W/L increases, the transistor current increases (equation 5), and the value of r_0 decreases ($r_0=1/\lambda I_{DS}$). The mn$_1$ I_{DS} equals $I(R_1)$ (Figure 60211).

Figure 60211 Mirror currents for L=0.9um, W = 4.5, 9, 18μm, R$_1$=30K

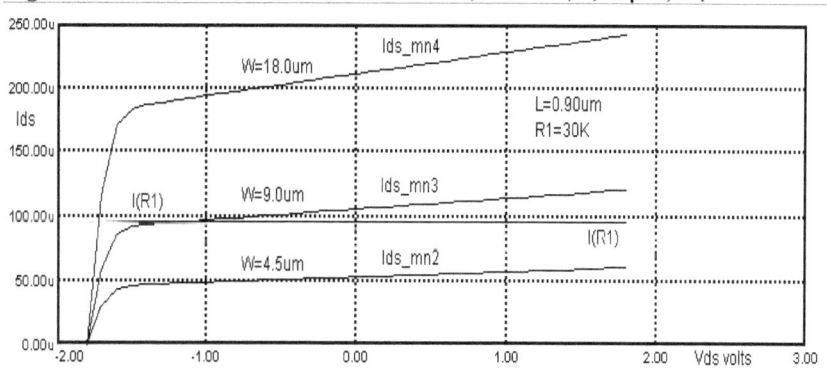

The MOS transistor must be operating in the saturation region to have high r_0. This means any design process must verify that node 3 output transistor voltage V_{DS} exceeds $V_{DSsat}=V_{GS}-V_{Tn}$ during circuit operation.

Problem 602 Reference equation 3. Calculate V_{GS} and V_{DSsat} for W/L= 10 and 1 where $\mu_n C_{ox}=150\mu A/V^2$, $I_{DS}=10uA$ and $V_T=0.5V$.

Select W/L=1 for mn₁ and W/L=0.5, 1, 2 for mn₂, mn₃, mn₄ to show the effect of W/L on r₀. Let $L=5L_{min}=10\lambda=0.9\mu m$ as in Figure 60211. However a reduction in W by a factor of ten (from W in figure 60211) shows a significant increase in r_0 predicted by equation 8b (Figure 60221).

Figure 60221 Mirror currents, for L=0.9um, W = 0.45, 0.9, 1.8μm, R₁=24K

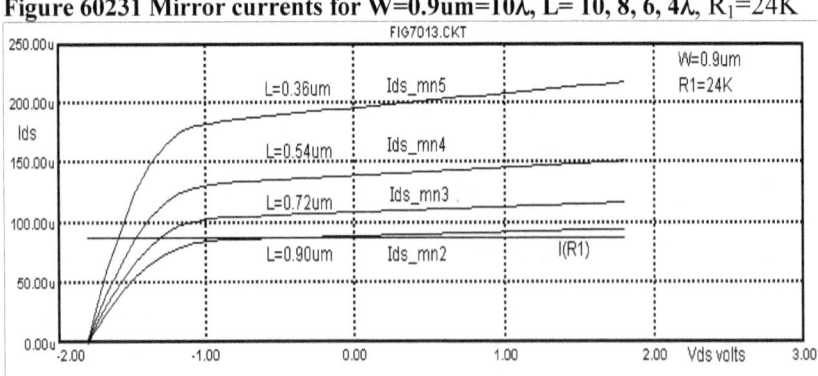

Select W=0.9um=10λ, and L= 10, 8, 6, 4λ Since $r_0=kL/W$ (equation 8b) r_0 decreases when L decreases (Figure 60231).

$$(8a) \quad I_{DS\lambda} = \frac{1}{2}\mu_n C_{ox}\frac{W}{L}(V_{GS}-V_T)^2\,(1+\lambda V_{DS}) \qquad (V_{DS} \geq V_{GS}-V_T > 0)$$

$$(8b) \quad \frac{1}{r_0} = \left(\frac{\partial I_{DS}}{\partial V_{DS}}\right)^{-1} = \frac{1}{2}\mu_n C_{ox}\frac{W}{L}(V_{GS}-V_T)^2\,(0+\lambda) \;\Rightarrow\; r_0 = \frac{1}{\lambda I_{DS}} = k\frac{L}{W}$$

Figure 60231 Mirror currents for W=0.9um=10λ, L= 10, 8, 6, 4λ, R₁=24K

```
Fig6021.ckt   nmos mirror, L constant, W varied, R₁=30K
.include 180_N1P1.txt
Vdd 1 0 DC 1.8
Vss 4 0 DC -1.8
R1 1 2 30K
mn1 2 2 4 4 N1 L=0.9u W=9.00u
mn2 3 2 4 4 N1 L=0.9u W=4.50u
mn3 3 2 4 4 N1 L=0.9u W=9.00u
mn4 3 2 4 4 N1 L=0.9u W=18.0u
V3 3 0 DC 0                        ; V3=Vds
.DC LIN V3 -1.8 1.8 0.1
.TEMP 27
.PLOT DC I(R1) ID(MN2) ID(MN3) ID(MN4) 0,250U
.PRINT DC V(2) V(3)
.PRINT DC I(R1) ID(MN2) ID(MN3) ID(MN4)
.end
```

```
Fig6022.ckt   nmos mirror, L constant, W varied, R₁=24K
.include 180_N1P1.txt
Vdd 1 0 DC 1.8
Vss 4 0 DC -1.8
R1 1 2 24K
mn1 2 2 4 4 N1 L=0.9u W=0.90u
mn2 3 2 4 4 N1 L=0.9u W=0.45u
mn3 3 2 4 4 N1 L=0.9u W=0.90u
mn4 3 2 4 4 N1 L=0.9u W=1.80u
V3 3 0 DC 0                        ; V3=Vds
.DC LIN V3 -1.8 1.8 0.1
.TEMP 27
.PLOT DC I(R1) ID(MN2) ID(MN3) ID(MN4) 0,250U
.PRINT DC I(R1) ID(MN2) ID(MN3) ID(MN4)
.end
```

```
Fig6023.ckt   nmos mirror, W constant, L varied, R₁=24K
.include 180_N1P1.txt
Vdd 1 0 DC 1.8
Vss 4 0 DC -1.8
R1 1 2 24K
mn1 2 2 4 4 N1 L=0.90u W=0.90u
mn2 3 2 4 4 N1 L=0.90u W=0.90u
mn3 3 2 4 4 N1 L=0.72u W=0.90u
mn4 3 2 4 4 N1 L=0.54u W=0.90u
mn5 3 2 4 4 N1 L=0.36u W=0.90u
V3  3 0 DC 0                       ; V3=Vds
.DC V3 -1.8 1.8 0.1
.TEMP 27
.PLOT DC I(R1) (ID(MN2)) (ID(MN3)) (ID(MN4)) (ID(MN5)) 0,250U
.PRINT DC I(R1) (ID(MN2)) (ID(MN3)) (ID(MN4)) (ID(MN5))
.end
```

Nmos replaces current mirror resistor

An MOS transistor vi constraint is essentially linear as it leaves I_{DS}, V_{DS} 0,0 coordinates. In this triode region the MOS transistor is a resistor that can replace the large area current mirror resistor. The design equations are straightforward, however there is a problem. *We do not know the exact MOS parameter values. We only have estimates. The design equation is*

Figure 605 Differential Amp

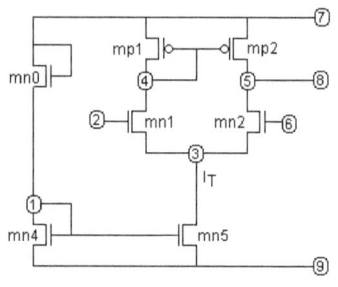

$$I_{DS} = \frac{1}{2}(\mu_n C_{ox})\frac{W}{L}(V_{GS} - V_{TN})^2 = \frac{1}{2}\beta(V_{GS} - V_{TN})^2 \quad (V_{DS} \ge V_{GS} - V_{TN} > 0)$$

We recycled Spice program 6051 (page 136) using different mn_0 W/L to get W/L=3/10, ID_{mn0}=54µA, and V_{GS}=2.78V. Now we can calculate $\mu_n C_{ox}$ (our initial guess from Figure M302 page 140 was 100µA/V^2)

$$\mu_n C_{ox} = \frac{L}{W}\frac{2I_{DS}}{(V_{GS} - V_{TN})^2} = \frac{10\lambda}{3\lambda}\frac{2 \cdot 54\,\mu A}{(2.78 - 0.7)^2 V^2} = 83.2\mu A/V^2$$

DC Operating Point Voltages

Node	Voltage	Node	Voltage	Node	Voltage	Node	Voltage
7	1.800	9	-1.800	6	0.000	2	0.000
1	−979.764	3	−801.450m	4	1.068m	5	1.068m

FIG6051.CKT	MN0	MN4	MN5
Model	N1	N1	N1
ID	54.072u	54.072u	16.328u
VGS	2.78	820.236m	820.326m
VDS	2.78	820.326m	998.550m
VBS	820.236m	0.000	0.000
VTH	662.536m	526.347m	526.036m
VDSAT	1.574	250.372m	250.576m
GM	25.087u	288.718u	86.366u
GDS	1.035u	8.111u	2.371u
GMB	15.540u	91.469u	27.402u

Problem 603 Reference *Nmos replaces current mirror resistor.* Find $\mu_n C_{ox}$ from the mn_0 data (W/L=3/10). Repeat for mn_4, mn_5.

Current Mirror Layout

The L in mirror transistors can be any value as set by the design.

Figure 601 Current Mirror

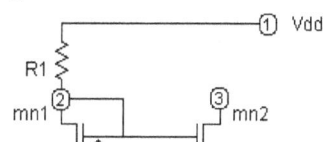

Figure 603 Current Mirror Layout (Figure 601), 1λ grid dots

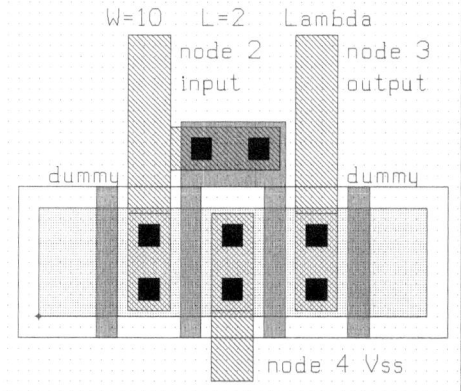

Two mn1 transistors make up the current mirror. Dummy gates enhance the transistor's match by minimizing undercutting the mn_1, mn_2 poly.

Figure 604 Current Mirror Layout (Figure 601), 1λ grid dots

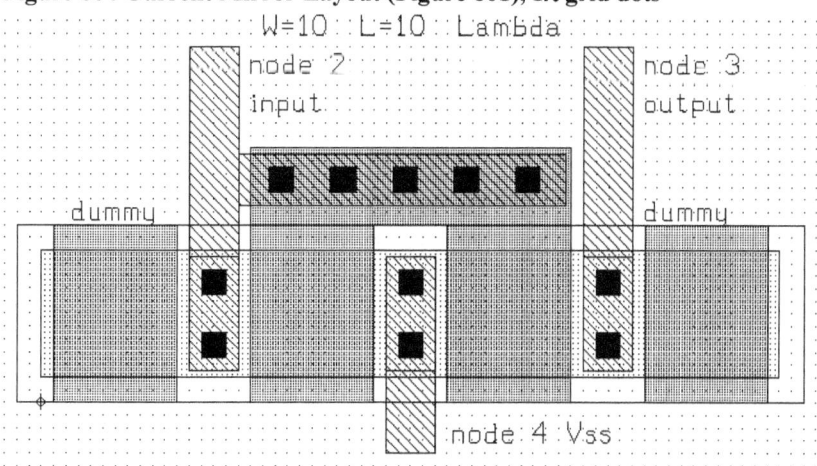

Problem 604 Draw the layouts in Figures 603, 604.

6.2 Differential Amplifier

A practical differential amplifier has high impedance inputs centered at 0 volts. Note: circuit 605 needs an output circuit such as a rail to rail inverter (e.g. Figure 607 page 140), with high impedance input and a low impedance output centered at 0V.

Transistors mn_1 and mn_2 are the basic differential amplifier pair with active current mirror load mp_1, mp_2 (Figure 605). Current mirror mn_0, mn_4, mn_5, biases the circuit (see *nmos replaces current mirror resistor* on page 130).

Specifications

(9) $slew\ rate = \dfrac{2V}{5ns}$

$C_{load} = 20\,fF$

$f_{-3Db} = 10MHz$

$gain\ T(0) = 50\ single\ ended$

$common\ mode - 0.6V\ to\ 1.3V$

Figure 605 Differential Pair

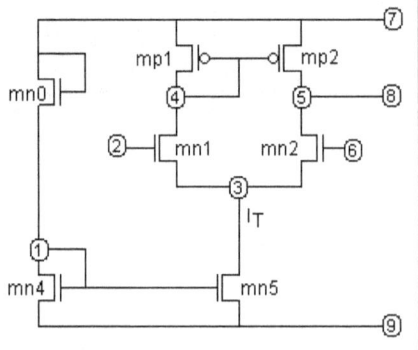

Differential amplifier design procedure The design problem is that we do not have accurate numbers for I_{DS}, V_{GS}, V_{TH}, μC_{ox} to use in I_{DS} equation 11 (page 133), which produces W/L numbers.

We start by estimating V_{GSn} and V_{GSp} to get initial μ_n and μ_p numbers for transistors in the circuit from Figures M302, M303 page 140. $V_T=V_P=0.4V$ in the figures. Increasing V_T to 0.8V decreases μ by less than 15%, which is why we use the figures for any V_T. Furthermore these are short channel numbers, which change as we recycle the design process.

Then we select I_{DS}, and estimate threshold voltages V_{THn} and V_{THp}, and use equation 11 on page 133 to calculate initial transistor W/L numbers.

Using the initial W/L numbers we run a Spice program that produces accurate I_{DS}, V_{GS}, and V_{TH} numbers (for that circuit) that are available in the Spice numeric output. Equation 11 then yields more accurate μC_{ox} numbers.

If the performance is not what we want, we use the accurate spice numeric output numbers to revise W/L numbers, run Spice, and check the numeric output again. If necessary we change our selections again, run Spice again, and check numeric output. We repeat the design process until we reach our goals.

The good news is that long nmos and pmos produce better analog circuit performance. Good news, because the *long* MOS I_{DS} equations are straightforward to use compared to the *short* MOS I_{DS} equations (page 44). We will use L=4λ and L=10λ.

I_{DS} **equation modified to include finite output resistance** Plots 60211, 60221, 60231 on pages 127, 128 show mos transistors have a finite output resistance r_0. The following modification includes r_0 in the I_{DS} equation.

$$(10) \quad g_0 = \frac{1}{r_0} = \frac{\partial I_{DS}}{\partial V_{DS}} = \frac{W\mu_n C_{ox}}{2L_{pinch}} (V_{GS} - V_T)^2 \frac{dL_{pinch}}{dV_{DS}} = I_{DS} \frac{dL_{pinch}}{L_{pinch} \, dV_{DS}} = \lambda I_{DS}$$

$$(11) \quad so \ that \ I_{DS} = \frac{1}{2}(\mu_n C_{ox})\frac{W}{L}(V_{GS} - V_{TN})^2(1+\lambda V_{DS})$$

where λ is the symbol for channel length modulation due to pinch off.[1] (This is not the layout λ.)

Gain and AC Transfer Function with Active Load The gain equals $g_m r_{output} = g_m r_0$ where

$$(12) \quad g_{m\_long} = \frac{\partial I_{DS}}{\partial V_{GS}} = \beta(V_{GS} - V_T) = \beta\sqrt{\frac{2I_{DS}}{\beta}} = \sqrt{2\beta I_{DS}} = \sqrt{2\frac{W\mu_n C_{ox}}{L} I_{DS}}$$

The output voltage v_{out} is i_{out} times the output impedance r_{out} in parallel with any load impedance Z_L. The amplifier's r_{out} equals the output impedances of transistors mn₂ and mp₂ in parallel. Design for gain A of 50 produces value 4 for W/L.

$$(13a) \quad A = \frac{v_{out}}{\Delta v_6} = \frac{r_{out} \cdot i_{out}}{\Delta v_6} = r_{out} \cdot g_{m2}\Delta v_6 \cdot \frac{1}{\Delta v_6} = g_{m2}\frac{r_{op} r_{on}}{r_{op} + r_{on}}$$

$$(13b) \quad A = g_{m2}\frac{\frac{1}{\lambda_p I_{DSp}}\frac{1}{\lambda_n I_{DSn}}}{\frac{1}{\lambda_p I_{DSp}} + \frac{1}{\lambda_n I_{DSn}}} = \frac{g_{m2}}{I_{DS}}\cdot\frac{1}{(\lambda_p + \lambda_n)} = \frac{1}{(\lambda_p + \lambda_n)}\sqrt{\frac{2\mu C_{ox}}{I_{DS}}\frac{W_2}{L_2}}$$

$$(13c) \quad \frac{W_2}{L_2} = [A(\lambda_p + \lambda_n)]^2 \frac{I_{DS}}{2\mu C_{ox}} = \left[50(0.06+0.04)\frac{1}{V}\right]^2 \frac{30\mu A}{2\cdot 100\mu A/V^2} = 3.75 \approx 4$$

---

[1] Gray et al, *Analysis and Design of Analog Integrated Circuits* page 43

Common Mode Range - Min and Max

There is a range of common mode input voltages over which the amplifier will operate as designed. I.e. all the transistors are in the saturation mode. The range is referred to as the common mode range (CMR). Equation 1b page 124 is the relevant equation for p or n transistors.

$$(1b) \ I_{DSsat} = \frac{1}{2}\mu C_{ox} \frac{W}{L}(V_{GS} - V_T)^2 = \frac{1}{2}\mu C_{ox} \frac{W}{L}V_{DSsat}^2$$

$$V_{DSsat} = \sqrt{\frac{2I_{DSsat}L}{(\mu C_{ox})W}} \quad \Rightarrow \quad \frac{W}{L} = \frac{2I_{DSsat}}{\mu C_{ox}V_{DSsat}^2}$$

Min The minimum common mode voltage V_{min} is reached when the current source transistor mn_5 (Figure 605) is about to enter the triode region where $V_{DSsat} = V_{GS} - V_{TN}$. Specify V_{min} in order to be able to calculate the W/L of the current source transistor.

$$V_{min} = V_{GS\_mn1} + V_{DSsat\_mn5} + V_{ss}$$

$$V_{DSsat\_mn5} = V_{min} - V_{ss} - V_{GS\_mn1}$$

$$V_{DSsat\_mn5} = V_{min} - V_{ss} - V_{TN} - \sqrt{\frac{2I_{DS}L_1}{(\mu C_{ox})W_1}}$$

And given V_{DSat_mn5} calculate $\dfrac{W_{mn5}}{L_{mn5}} = \dfrac{2I_{DS}}{\mu C_{ox}V_{DSsat\_mn5}^2}$

Max The maximum voltage is reached when transistors mn_1 and mn_2 (Figure 605) are about to enter the triode region where $V_{DSsat} = V_{GS} - V_{TN}$. Specify V_{max} in order to be able to calculate the W/L of the active load pmos current mirror transistors.

$$V_{max} = V_{dd} - V_{DS\_mp1} - (V_{DS\_mn1\_sat}) + V_{GS\_mn1}$$
$$= V_{dd} - V_{GS\_mp1} - (V_{GS\_mn1} - V_{TN}) + V_{GS\_mn1}$$
$$= V_{dd} - V_{GS\_mp1} + V_{TN}$$

$$V_{GS\_mp1} = V_{dd} + V_{TN} - V_{max}$$

Now that we know V_{GS_mp1} we can calculate W/L from the equation

$$(V_{GS\_mp1} - V_{TP})^2 = \frac{2I_{DSsat}L}{(\mu C_{ox})W}$$

$$\frac{W_{mp1}}{L_{mp1}} = \frac{W_{mp2}}{L_{mp2}} = \frac{2I_{DS}}{\mu C_{ox}(V_{GS\_mp1} - V_{TP})^2}$$

Differential amplifier design procedure continued
1. Slew rate determines minimum I_T.

$$(14) \quad i = C \frac{dv}{dt} \quad \Rightarrow \quad I_T \geq 20\,fF \frac{2V}{5ns} = 8\mu A$$

2.mn_0 Supply voltage is 1.8V+1.8V=3.6V. Assume μC_{ox}=100$\mu A/V^2$, I_{DS}=60μA, V_{GS}=2.7V, V_{TH}=0.7V. This leaves 0.9V for mn_4, which the $V_{GS}-V_{TH}$ difference will not make W_4/L_4 too large. Make mn_0 long by letting L=10λ.

$$(15) \quad \frac{W_{mn0}}{L_{mn0}} = \frac{2I_{DS}}{\mu C_{ox}(V_{GS}-V_{TH})^2} = \frac{2 \cdot 60\mu A}{100\mu A/V^2 \cdot (2.7-0.7)^2} = \frac{1.2}{4} = 0.3 = \frac{3\lambda}{10\lambda}$$

Spice program 6051 data shows that mn_0 I_{DS}=54.072uA, V_{GS}=2.780V.

3. mn_4 $V_{GSmn4}=V_{GSmn5}=V_7-V_9-V_{GSmn0}$=1.8+1.8–2.7=0.9V.
When t_{ox}=4nm, μ=200 cm^2/Vs (Fig M302, page 140) so that
μC_{ox}=200 cm^2/Vs $\times$ 8.63fF/μm^2 = 173 $\mu A/V^2$. Assume V_{TH}=0.5V. Make
mn_4 long by letting L=4λ.

$$(16) \quad \frac{W_4}{L_4} = \frac{2I_{DS_4}}{\mu_n C_{ox}(V_{GS4}-V_{THn})^2} = \frac{2 \cdot 60\mu A}{173\mu A/V^2(0.9-0.5)^2 V^2} = 4.34 = \frac{17\lambda}{4\lambda}$$

4. mn_5 Reduce I_{DS5} to increase gain.

$$(17) \quad \frac{W_5}{L_5} = \frac{I_{DS5}}{I_{DS4}} \cdot \frac{W_4}{L_4} = \frac{20\mu A}{60\mu A} \cdot \frac{17}{4} \approx \frac{5\lambda}{4\lambda}$$

The gain is sensitive to the I_T=I_{DS} value. Spice shows T–54.29. Compare to 50. Varying the mn_5 W is instructive. Try W=0.90u, 0.56u, 0.45u etc.

5. mp1, mp2 Choose maximum common mode voltage as 1.5V. And for V_{GS}=0.75, μC_{ox}=80cm^2/Vs $\times$ 8.63fF/μm^2=69$\mu A/V^2$ from Figure M303.

$$V_{GS\_mp1} = V_{dd} + V_{TN} - V_{max} = 1.8 + 0.45 - 1.5 = 0.75V$$

$$(18) \quad \frac{W_{mp1}}{L_{mp1}} = \frac{W_{mp2}}{L_{mp2}} = \frac{2I_{DS}}{\mu C_{ox}(V_{GS\_mp1}-V_{TP})^2} = \frac{2 \times 10\mu A}{69\mu A/V^2(0.75-0.45)^2 V^2} = 3.22 \approx 4 = \frac{40\lambda}{10\lambda}$$

6. mn1, mn2 V_{GS}=0.80V $\rightarrow$ μC_{ox}=200 $\mu A/V^2 \times 0.863$=173

$$(19) \quad \frac{W_{mn1}}{L_{mn1}} = \frac{W_{mn2}}{L_{mn1}} = \frac{2I_{DS}}{\mu_n C_{ox}(V_{GS\_mn1}-V_{TN})^2}$$

$$= \frac{2 \cdot 10\mu A}{173\mu A/V^2(0.80-0.60)^2 V^2} = 2.89 \approx 4 = \frac{40\lambda}{10\lambda}$$

Spice program 6051 Amplifier L=0.90☐m, Mirror L=0.90☐m

Fig6051.ckt cmos diff amp L=0.90u=5Lmin

.include 180_N1P1.txt

V7 7 0 DC 1.8
V9 9 0 DC -1.8

V6 6 0 DC 0 AC 1   sin(0 0.005 1e6 0 0)

R2 2 0 1K ;1E10

mn0 7 7 1 9 N1 L=0.90u W=0.27u   ;W/L=3/10
mn4 1 1 9 9 N1 L=0.36u W=1.53u   ;W/L=17/4
mn5 3 1 9 9 N1 L=0.36u W=0.45u   ;W/L=5/4

mp1 4 4 7 7 P1 L=0.90u W=3.60u   ;W/L=40/10
mp2 5 4 7 7 P1 L=0.90u W=3.60u
mn1 4 2 3 9 N1 L=0.90u W=3.60u
mn2 5 6 3 9 N1 L=0.90u W=3.60u

*.PLOT AC VP(4) VP(5) -200,50
.DC V6 -0.5 0.5 0.01
.PLOT DC V(6) V(4) V(5) -2.5,2.5

.TRAN 1e-008 1e-005 0 1e-008
.PLOT TRAN (50*V(6)) V(5) -0.5,1.5
.AC DEC 25 1e+006 1e+010

.TEMP 27
.PLOT AC VDB(4) VDB(5) -50,50
.PRINT AC V(5)
.PRINT AC VDB(4) VDB(5)
.end

DC Operating Point Voltages
Node Voltage Node Voltage Node Voltage

Node	Voltage	Node	Voltage	Node	Voltage
7	1.800	9	-1.800	6	0.000
2	0.000	1	-979.764m	3	-801.450m
4	1.068	5	1.068		

Gain

GHz	DB(V(4))	DB(V(5))	V(5)
0.001	-5.506	34.686	54.239

Figure 60511 Frequency Response

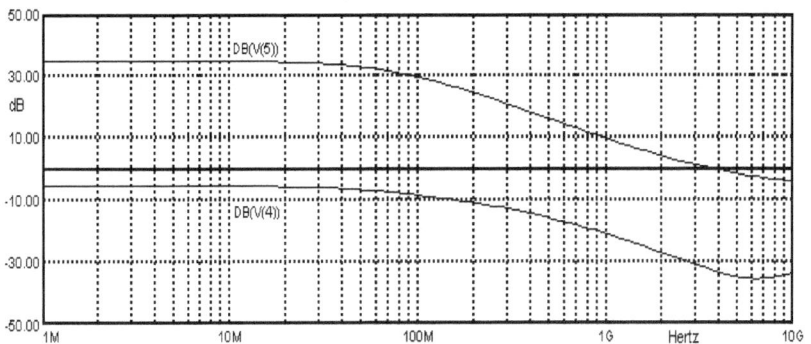

Figure 60512 DC Transfer Function

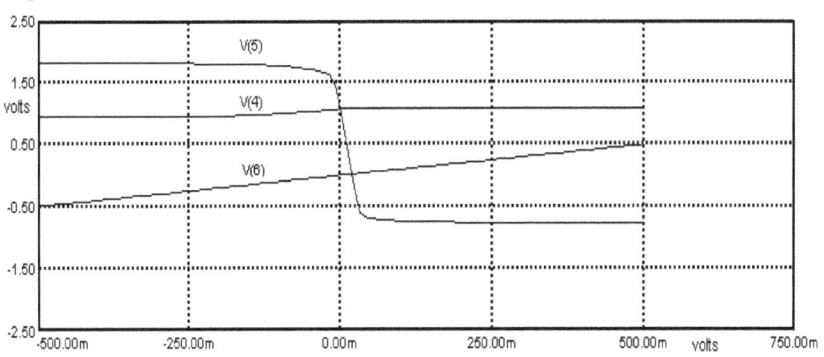

Figure 60513   1MHz Sine Wave Transient Response, Gain=54

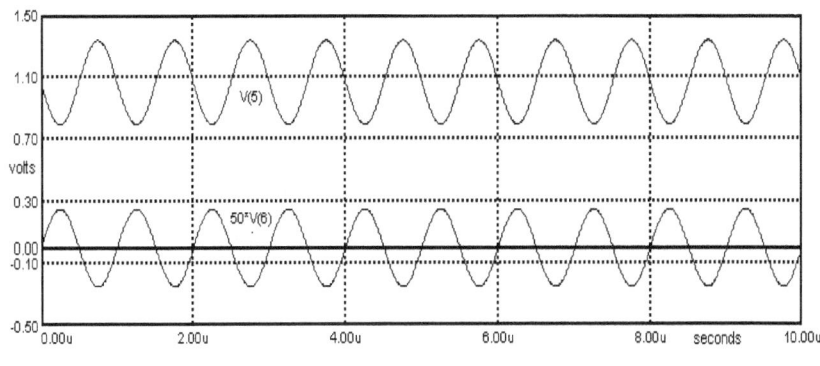

Differential Amplifier Layout

Spice Program 6051 uses
```
mn0 7 7 1 9 N1
+ L=0.90u W=0.27u    ;W/L=3/10

mn4 1 1 9 9 N1
+ L=0.36u    W=1.53u      ;W/L=17/4
*Ids=57uA

mn5 3 1 9 9 N1
+ L=0.36u    W=0.45u      ;W/L=5/4
*IT=2Ids=104uA

mp1 4 4 7 7 P1
+ L=0.90u    W=3.60u          ;W/L=40/10
mp2 5 4 7 7 P1 L=0.90u W=3.60u

mn1 4 2 3 9 N1 L=0.90u    W=3.60u        ;W/L=40/10
mn2 5 6 3 9 N1 L=0.90u    W=3.60u
```

Figure 605 Amplifier Circuit

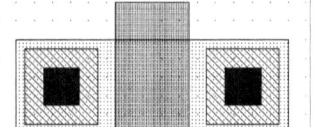

Figure 606a mn$_0$ L=10λ, W=3λ (1λ dots)

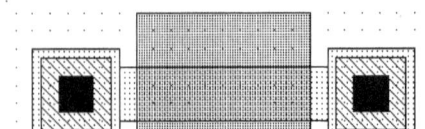

Figure 606c mn$_5$ L=4λ, W=5λ

Figure 606b mn$_4$ L=4λ, W=17λ (1λ dots)

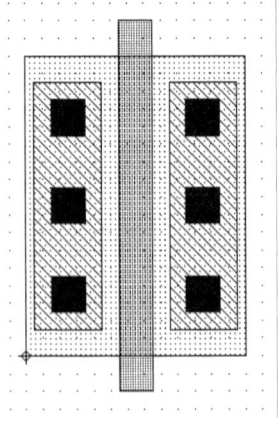

Problem 605 Draw the layouts in Figures 606a, 606b, 606c, 606d, 606e.

Figure 606d mn$_1$, mn$_2$, mp$_1$, mp$_2$, L=10λ, W=20λ (1λ dots)

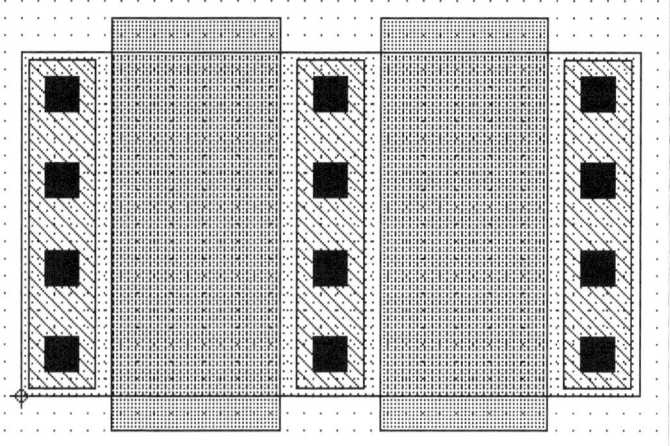

Figure 606e Differential Amplifier Layout (2λ dots)

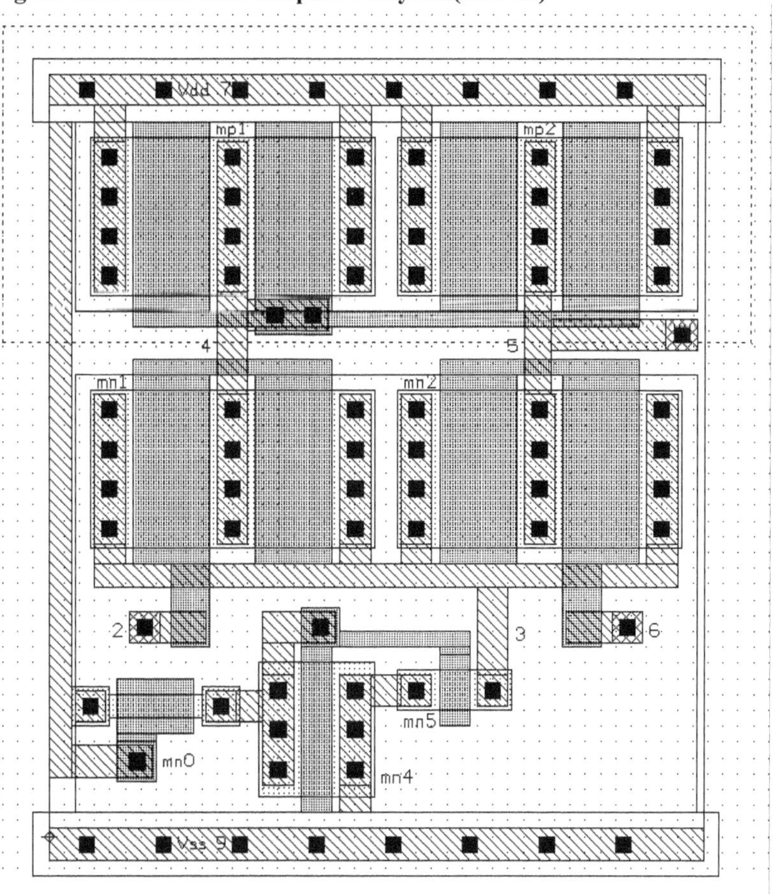

6.3 Operational Amplifier

Operational amplifiers have high gain and low output impedance. These properties are realized by current mirrors (mn_0, mn_4, mn_5, mn_6), a differential amplifier (mp_1, mp_2, mn_1, mn_2), a second gain stage (mp_3), and an inverter (mp_4, mn_3). To reduce chip area the current mirror resistor is replaced by nmos transistor mn_0.

Figure 607 Operational Amplifier

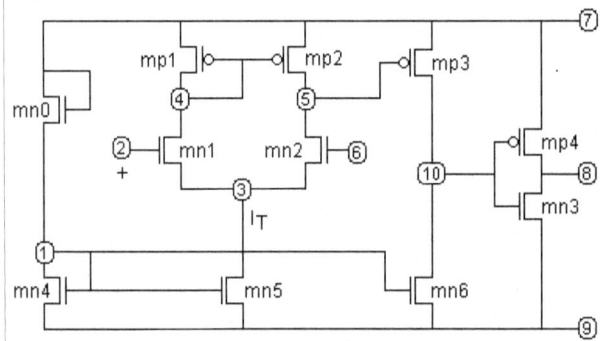

Figure M302 Effective electron mobility cm²/Vs

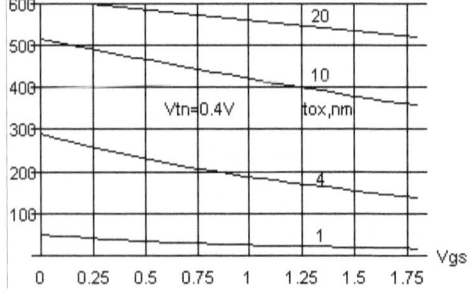

Figure M303 Effective hole mobility cm²/Vs

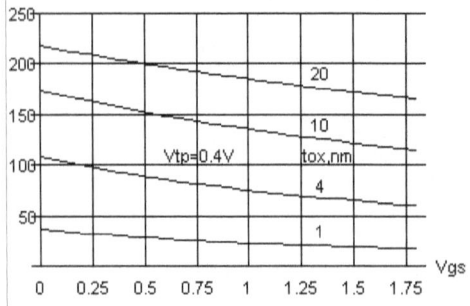

Design procedure This design is the result of several iterations confirmed by Spice, which we recommend you do for every design to ascertain how sensitive the design is to W/L values. The μC_{ox} value for a V_{GS} value is found in Figures M302, M303. Revised values are found in Spice numeric data.

1. mn$_0$
2. mn$_4$
3. mn$_5$
4. mp$_1$, mp$_2$
5. mn$_1$, mn$_2$
Same as design steps 2 to 6 on page 135.

6. mn$_6$ After several recycles W/L=4/10.

(20) $I_{DS} = \dfrac{1}{2}(\mu_n C_{ox})\dfrac{W}{L}(V_{GS} - V_{TN})^2$

(21) $8\mu A = \dfrac{1}{2}\left(173\dfrac{\mu A}{V^2}\right)\dfrac{W}{L}(0.80 - 0.45)^2 V^2 \quad \Rightarrow \quad \dfrac{W}{L} = 0.75 \approx 0.7 = \dfrac{7\lambda}{10\lambda}$

7. mp$_3$ Recycled to L=0.45u, W=2.25u, W/L=25/5

(22) $I_{DS} = \dfrac{1}{2}(\mu_n C_{ox})\dfrac{W}{L}(V_{GS} - V_{TN})^2$

(23) $8\mu A = \dfrac{1}{2}(69)\dfrac{W_{mp3}}{L}(0.65 - 0.45)^2 \quad \Rightarrow \quad \dfrac{W_{mp3}}{L} = 5.8 \approx 6 = \dfrac{30\lambda}{5\lambda}$

8. mn$_3$, mp$_4$ Let the push-pull inverter source 2.5mA. Guess V_{GS}=1.85V. Then μ_n=150, and $\mu_n C_{ox}$=130 (Figure M302).

(24) $I_{DS} = \dfrac{1}{2}(\mu_n C_{ox})\dfrac{W}{L}(V_{GS} - V_{TN})^2$

(25) $2500\mu A = \dfrac{1}{2}(130)\dfrac{W_{mn3}}{L}(1.75 - 0.45)^2 \Rightarrow \dfrac{W_{mn3}}{L} = 23 \approx 20 = \dfrac{40\lambda}{2\lambda} = \dfrac{3.60\mu m}{0.18\mu m}$

(26) $\dfrac{W_{mp4}}{L_{min}} = \dfrac{\mu_n C_{ox}}{\mu_p C_{ox}}\dfrac{W_{mn3}}{L_{min}} = \dfrac{130}{69}40 = 75 \approx 80$

Problem 606 Reference Figure 607 and Spice program 6071. Run Fig6071.ckt. Use mp$_4$ and mn$_3$ G$_{DS}$ data to show that r$_{out}$=590Ω.

Figure 60711 Frequency Response

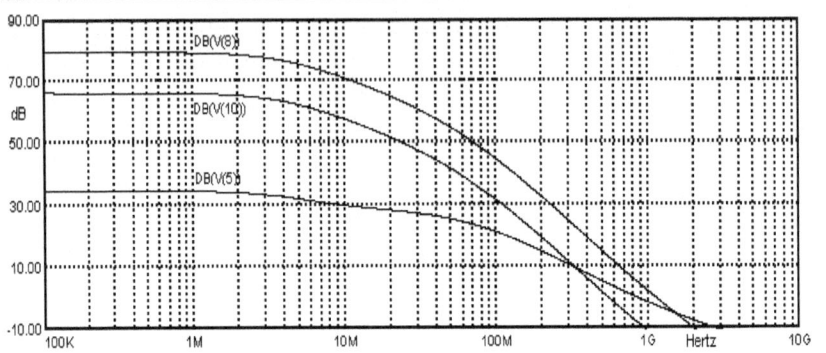

Figure 60712 DC Transfer Function

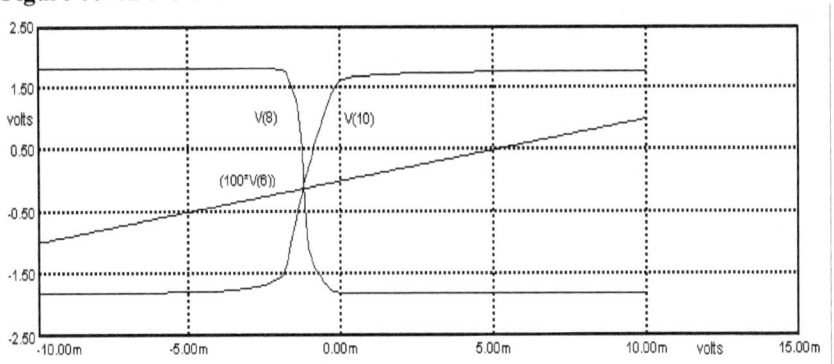

Figure 60713 Sine Wave Transient Response, Gain=9,200

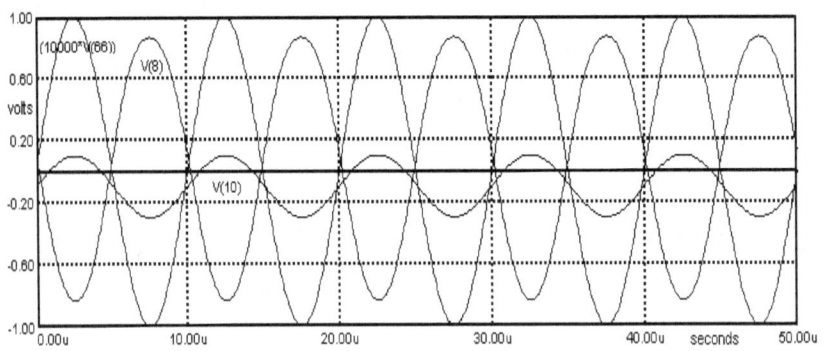

Spice Program 6071

Fig6071.ckt cmos diff amp L=0.90u and Loutput=0.18u

.include 180_N1P1.txt
.lib G:\!B\NLPbooks\ee105pdf\cm06spice\mos018_2.lib
.lib G:\!B\NLPbooks\ee105pdf\cm06spice\mos018_L.lib

V7 7 0 DC   1.8
V9 9 0 DC   −1.8
V66 66 0   DC 0   AC 1   sin(0 0.0001 1e5 0 0)
V6   6 66 DC −0.001225   ;bias offset
R2 2 0 1k

mn0 7 7 1 9 N1 L=0.90u W=0.27u   ;W/L=3/10
mn4 1 1 9 9 N1 L=0.36u W=1.53u   ;W/L=17/4
mn5 3 1 9 9 N1 L=0.36u W=0.45u   ;W/L=5/4

mp1 4 4 7 7 P1 L=0.90u W=3.60u   ;W/L=40/10
mp2 5 4 7 7 P1 L=0.90u W=3.60u
mn1 4 2 3 9 N1 L=0.90u W=3.60u
mn2 5 6 3 9 N1 L=0.90u W=3.60u

mp3 10 5 7 7 P1 L=0.45u W=2.25u   ;W/L=25/5
mn6 10 1 9 9 N1 L=0.90u W=0.36u   ;W/L=4/10

xmp4 8 10 7 7 mp11   ;W/L=80/2
xmn3 8 10 9 9 mn7   ;W/L=40/2

*.PLOT AC VP(3) VP(4) VP(5) -200,50
*.PRINT AC V(5) V(10) V(8) VDB(10) VDB(8)
*.PLOT AC VDB(5) VDB(10) VDB(8) VDB(3) -20,80
*.PRINT AC V(5) V(10) V(8)
.DC V6 -0.01 0.01 0.0001
.PLOT DC (100*V(6)) V(10) V(8) -2.5,2.5
.TRAN 1e-007 5e-005 0 1e-007
*.PLOT TRAN (10000*V(66)) V(10) V(8) -1,1
.AC DEC 1000 100000 1e+010
.TEMP 27
.PLOT AC VDB(5) VDB(10) VDB(8) -10,90
.PRINT AC V(3) V(5) V(10) V(8)
.PRINT AC VDB(5) VDB(10) VDB(8)
.end

Figure 607e Operational Amplifier Layout (2λ dots)

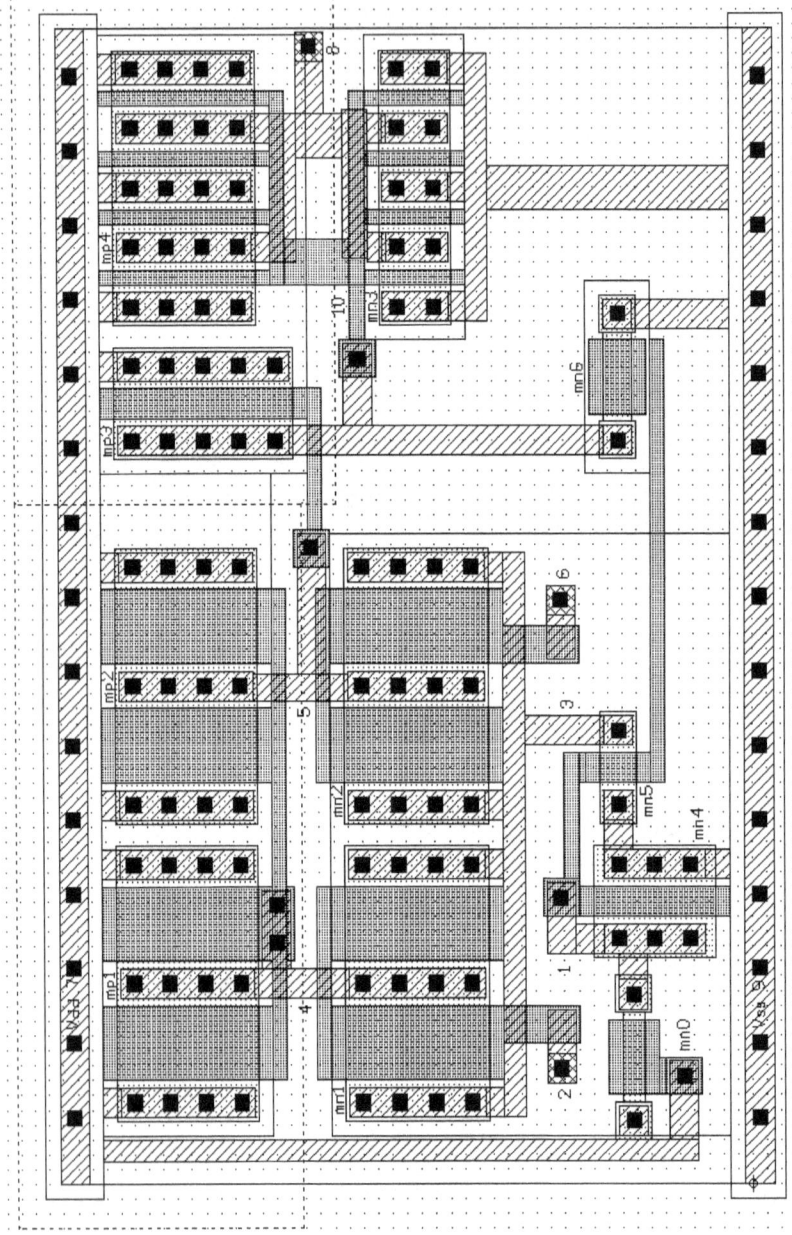

Problem 607 Draw the layout in Figure 607a.

Process Parameters ($L_{min} = 2\lambda = 0.18$ μm, $t_{ox} = 4.08$nm)

$$C_{ox} = \frac{\varepsilon_{SiO2}\varepsilon_0}{t_{ox}} = \frac{3.9 \times 8.85}{0.004} \frac{aF}{\mu m \times \mu m} = 8629 \frac{aF}{\mu m^2} = 8.63 \frac{fF}{\mu m^2}$$

$$C_{gate} = C_{ox}WL = C_{ox}\frac{W}{L}L^2 = C_{ox}L^2\frac{W}{L}$$

$$= 8.63\frac{fF}{\mu m^2}(0.18\mu m)^2\frac{W}{L_{min}} = 0.280\frac{W}{L_{min}}fF$$

Transistor drive current (C in fF)

nmos i_{drive}=0.61 mA/μm   pmos i_{drive}=0.26 mA/μm

$$\frac{W}{L} = \frac{1}{L}\frac{C_{load}}{i_{drive}}\frac{dv}{dt}$$

$$\frac{W_p}{L_{min}} = \frac{1}{0.18\mu m} \cdot \frac{C_{load}\ fF}{0.26mA/\mu m} \cdot \frac{1.8V}{100pS} = 0.3846C_{load}\ fF$$

$$\frac{W_n}{L_{min}} = \frac{1}{0.18\mu m} \cdot \frac{C_{load}\ fF}{0.60mA/\mu m} \cdot \frac{1.8V}{100pS} = 0.1667C_{load}\ fF$$

Resistance
$R_{metal\ k} - 0.07$ Ω/sq   (k = 1, 2, 3, 4, 5, 6) $R_{poly} = 7.8$ Ω/sq

Capacitance

Layer	aF/μm	line width	line sep μm	aF/λ	width λ
poly	144	0.16	0.27	13.0	2
metal 1	244	0.23	0.23	26.0	3
	110	0.23	2.00	11.6	3
metal 2	220	0.28	0.28	19.1	3
	85	0.28	2.00	7.4	3
metal 3,4,5	215	0.28	0.28	18.7	3
	78	0.28	2.00	6.8	3
metal 6	260	0.44	0.46	19.2	4
	110	0.44	2.00	8.1	4

6.4 Resistors

Analog circuits may use resistors, capacitors, and inductors in addition to mos transistors. Here we briefly explain how to layout R.

Units of materials used to make resistors are ρ ohms-meter and ρ/t ohms/square. Typical order-of-magnitude values for an integrated circuit process are as follows. The resistance of metal and poly layers used in an integrated circuit is

$R_{metal\ k} = 0.07\ \Omega/sq\ (k = 1\ to\ 6)$ $R_{poly} = 7.8\ \Omega/sq$ $R_{nwell} = 150\ \Omega/sq$ est.

In what follows L and W are length and width of a wire with rectangular cross-section $t \times W$. (L and W are **not** the mos transistor parameters.)

$(27a)\quad R(ohms) = \dfrac{resistivity \times length}{cross\ sectional\ area} = \dfrac{\rho L}{tW}$

the unit for resistivity is derived as follows :

$(27b)\quad \rho = Rt\dfrac{W}{L} \quad \Rightarrow \quad ohm \times meter \times \dfrac{meter}{meter} = ohm-meter$

The resistance of a *square* of material of any size in a CMOS layer is a constant. Here is why. Divide a piece of material into squares W units on a side. Then material of length L and width W is L/W squares (28a).

$(28a)\quad \dfrac{material\ area}{square} = \dfrac{LW}{W^2} = \dfrac{L}{W}\left(squares\ in\ a\ piece\ of\ metal\ L\ long \times W\ wide\right)$

$(28b)\quad R(ohms) = \dfrac{resistivity \times length}{thickness \times width} = \dfrac{\rho L}{tW} = \dfrac{\rho}{t}\dfrac{ohm-meters}{meters} \times \dfrac{L}{W}squares$

$(28c)\quad this\ implies\ the\ \dfrac{\rho}{t}\ dimensions\ are\ \dfrac{ohms}{square}$

For example, a rectangular poly wire, 4λ wide and 10000λ long ($0.36\mu m \times 900\mu m$), has 2500 squares ($10000/4$) and a total resistance of 19,500 ohms.

$(29)\quad R = \dfrac{\rho}{t} \times \dfrac{L}{W} = 7.8\dfrac{ohm}{square} \times \dfrac{10^4\lambda}{4\lambda}squares = 7.8 \times 2500 = 19,500\Omega$

Designing Resistors

In many circuits absolute resistor values are not the issue - resistor ratios are. *Use the same width* for all resistors on a chip so that all resistor values vary in the same ratio for different process runs. *Areas quoted do not include the poly to metal₁ pads.* R=50×4×7.8=1560 ohms.

Figure 608 Resistor in the poly layer - L·W=50λx4λ (1λ grid dots)

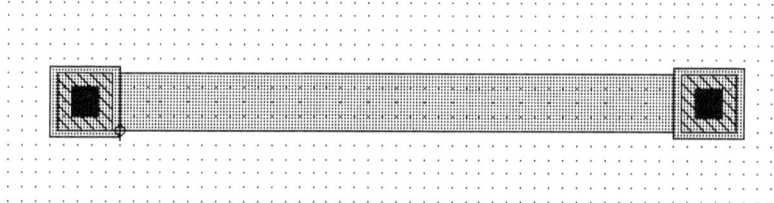

Concentrating a resistor area produces corners in the layout (Figure 609). Corner resistance is hard to evaluate. A reasonable approximation is to use the length of the poly centerline to the contact centers divided by the width. R=(4×50×4+36)7.8=6521 ohms.

Figure 609 R in the poly layer – 4·L·W+3·3·4=4·50λ·4λ+36λ² (1λ dots)

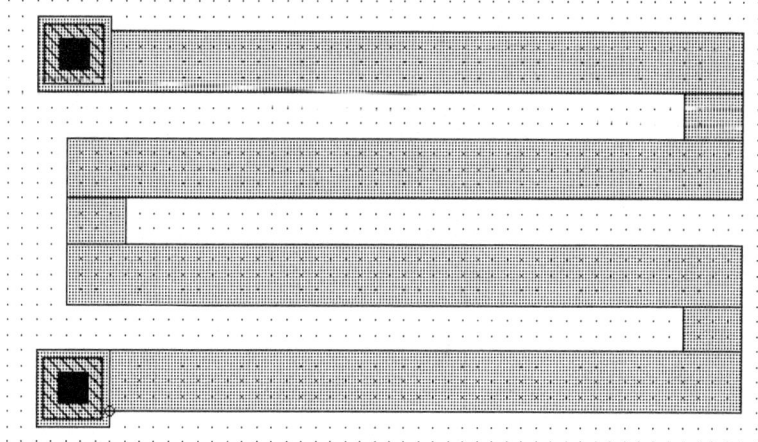

Notes: Poly is covered by a layer of conductive salicide whose resistance is in parallel with the poly resistance that produces R_{poly} = 7.8 Ω/sq. If a process option is available the salicide layer is removed and the stripped poly 7.8 Ω/sq increases to about 78 Ω/sq.

In some circuits resistor fingers need to be shielded from each other by grounding interlaced fingers (Figure 610).

Figure 610 Shielded Resistor, 1λ grid dots

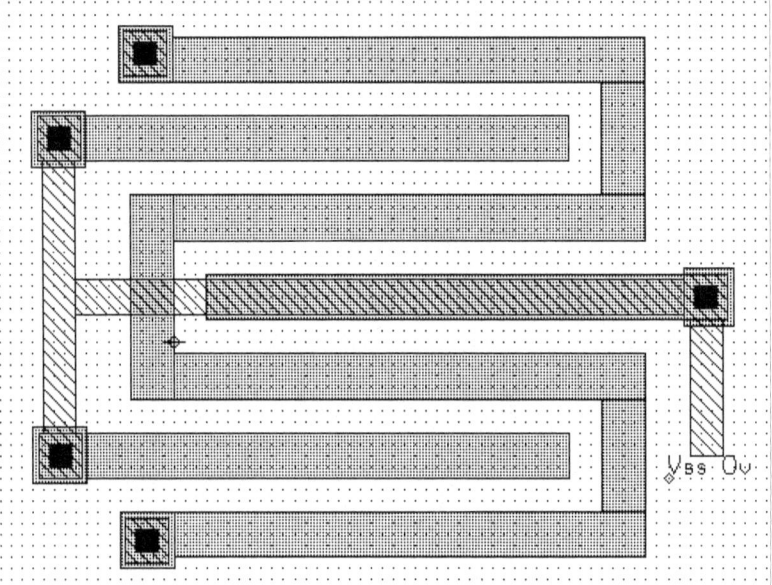

Balance is an issue in some sensitive circuits.

When two 180 degree out of phase signals drive two resistors the coupling effects due to parasitic capacitors are balanced by interlaced resistor layouts (Figure 611).

Figure 611 "balanced" Resistors, 1λ grid dots

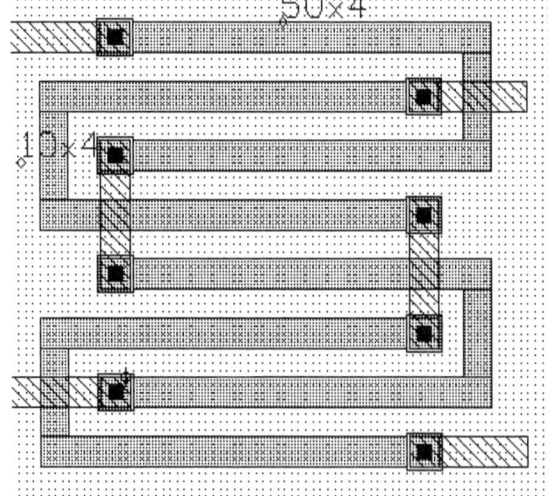

Problem 608 Draw the layouts in Figures 608, 609, 610, 611.

6.5 Capacitors

There several ways to create capacitors on a chip. The MOS gate capacitor is one way (Figure 612).

Figure 612 MOS Capacitors

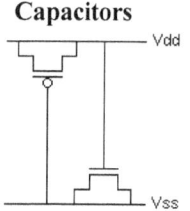

Process Parameters

$t_{ox} = 4.08\text{nm}$

$$(30) \quad C_{ox} = \frac{\varepsilon_{SiO2}\varepsilon_0}{t_{ox}} = \frac{3.9 \times 8.85}{0.004} \frac{aF}{\mu m \times \mu m}$$

$$= 8629\frac{aF}{\mu m^2} = 8.63\frac{fF}{\mu m^2}$$

for example

$$(31) \quad C_{gate} = C_{ox}WL = 8.63\frac{fF}{\mu m^2}(20\lambda \times 20\lambda)(0.09\frac{\mu m}{\lambda})^2 = 28\,fF$$

Figure 613 MOS Capacitors 28fF each - W×L=20×20 λ^2

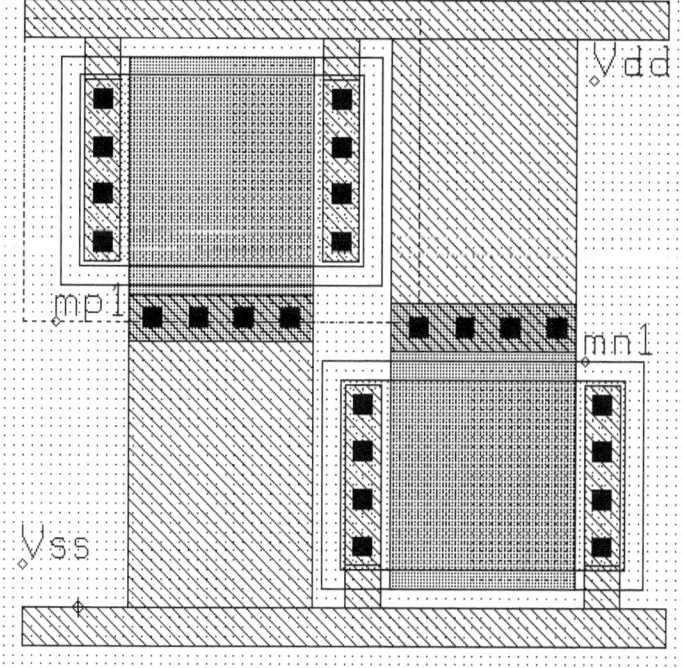

Another way to create capacitors is to use the poly and poly$_2$ layers as capacitor plates. The poly$_2$ layer is a second poly layer that is physically above the first poly layer (Figure 614).

Figure 614 Poly/Poly2 Capacitor with Contacts, 1λ grid dots

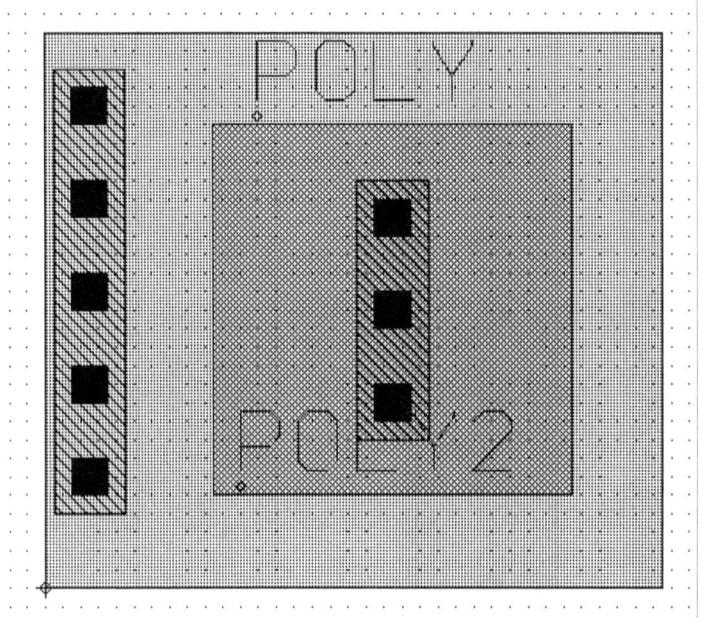

$$(32) \quad C_{pp} = \frac{\varepsilon_{SiO2}\varepsilon_0}{t_{pp}} = \frac{3.9 \times 8.85}{0.016} \frac{aF}{\mu m \times \mu m} = 2157 \frac{aF}{\mu m^2} = 2.16 \frac{fF}{\mu m^2}$$

poly2 is $20\lambda \times 20\lambda$

$$(33) \quad C_p = C_{pp}WL = 2.16 \frac{fF}{\mu m^2}(20\lambda \times 20\lambda)(0.09 \frac{\mu m}{\lambda})^2 = 7 fF$$

Rule	Poly$_2$	Lambda
11.1	Minimum width	7
11.2	Minimum spacing	3
11.3	Minimum poly overlap	5
11.4	Min spacing to active or well edge (not illustrated)	2
11.5	Min spacing to poly contact	6
11.6	Min spacing to unrelated metal	2

Problem 609 Draw the layouts in Figures 613, 614.

7 CMOS Circuit Design Fundamentals

7.1 Circuit Delay and Circuit C_{Load}

Figure 703 Inverter

Propagation delay is defined in product data sheets as t_{PHL} and t_{PLH} when inputs and outputs cross $V_{dd}/2$. In practice $t_{PHL}=t_{PLH}=t_p$ in almost every circuit. The immediate goal is an equation for the circuit delay t_p as a function of C_{Load}, where a basic time unit τ_0 is a characteristic of the semiconductor fabrication process being used. We will show that

(1) $t_p = \tau_0 + kF$ (*where F is the fanout and k is a constant*)

The propagation delay through any CMOS circuit increases as the capacitance load C_{Load} increases. This is illustrated by Spice Program 7031. Let the fanout $F=C_{Load}/C_{in}$ where C_{in}=18.3fF is the input C of one circuit (equation 5). From Figure 70312 the delay data are

(2)
F	0	1	2	3	4	5	6
delay ps	τ_0	47	60	71	81	91	104

Extrapolate the plot of propagation delay (Figure 704) back to the delay axis t_p where F = 0. The point of intersection is labeled τ_0, where τ_0 is the no load circuit delay. τ_0 represents the fact that in any semiconductor fabrication process every circuit has a no load delay.

Figure 704 Delay vs F=C_{load}/C_{in}

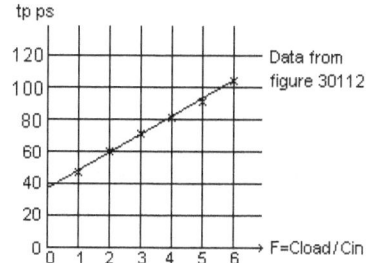

The plot (Figure 704) produces equation (1), which shows that delay is proportional to F. And, an analysis of short channel transient response produces equation (3) where τ_0 is independent of C_L, and dependent on the process parameter values.

$$(3)\quad \tau_0 = \frac{C_L}{v_{sat}WC_{ox}}$$

$$= \frac{LWC_{ox}}{v_{sat}WC_{ox}} = \frac{L}{v_{sat}}$$

Delay t_p equation (1) parameters are computed from the data follow.

(4a) $t_p = \tau_0 + kF$ $\rightarrow$ $36 = \tau_0 + 0$ and $104 = \tau_0 + 6k$ (2 data points)

(4b) $104 = 36 + 6k$ $\rightarrow$ $k = \dfrac{104-36}{6} = \dfrac{68}{6} = 11.3$

(4c) $t_p = \tau_0 + kF = 36 + 11.3 Fanout$ ps

From Figure 70311

(5) $C_{in} = i \dfrac{dt}{dv} = (0.110 + 0.055)mA \dfrac{200\,ps}{1.8V} = 18.3\,fF$

Figure 70311 Inverter gate currents

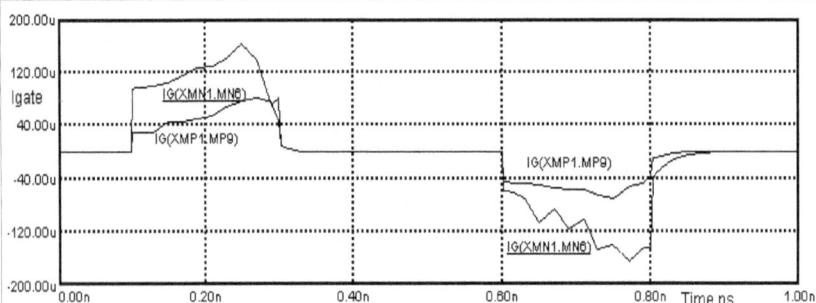

Figure 70312 Inverter Delay vs F=C_{load}/C_{in} (expanded time scale)

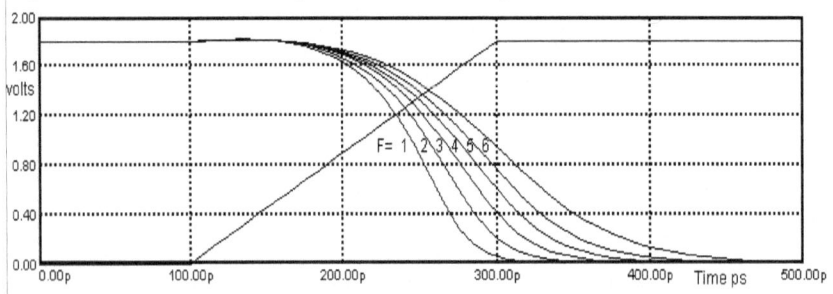

Figure 70313 Inverter Delay vs F=C_{load}/C_{in}

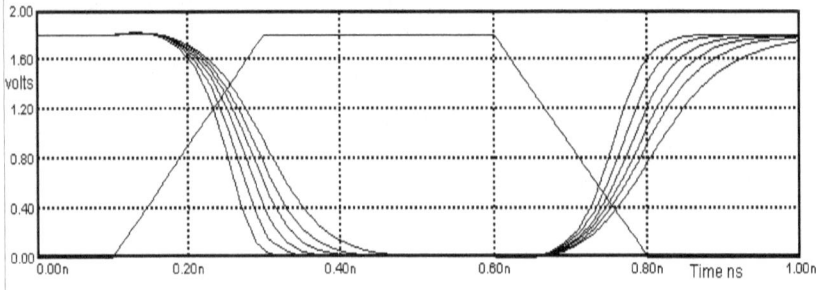

Spice program 7031

```
Fig7031.ckt   Inverter Delay
.include 180_N1P1.txt
.lib G:\!B\NLPbooks\ee105pdf\cm07spice\mos018.lib

Vdd 98 0 DC 1.8
V9 9 0 PULSE(0 1.8 100p 200p 200p 300p 1000p)

XMP1 1 9 98 98 mp9        ; W/L=60/2
XMN1 1 9 0   mn6          ; W/L=30/2
C1 1 0 18f

XMP2 2 9 98 98 mp9
XMN2 2 9 0 mn6
C2 2 0 36f

XMP3 3 9 98 98 mp9
XMN3 3 9 0 mn6
C3 3 0 54f

XMP4 4 9 98 98 mp9
XMN4 4 9 0 mn6
C4 4 0 72f

XMP5 5 9 98 98 mp9
XMN5 5 9 0 mn6
C5 5 0 90f

XMP6 6 9 98 98 mp9
XMN6 6 9 0 mn6
C6 6 0 108f

.AC DEC 25 1e+006 1e+008
.PLOT AC VDB(98) VP(98) VDB(9) 1500,-2250
*.PLOT TRAN IG(XMN1.MN6) IG(XMP1.MP9) -0.2M,0.2M
.TRAN 5e-012 1e-009 0
.TEMP 27
.PLOT TRAN V(1) V(2) V(3) V(4) V(5) V(6) V(9) 0,2
.end
```

7.2 Circuit Delay and Input Switching Time

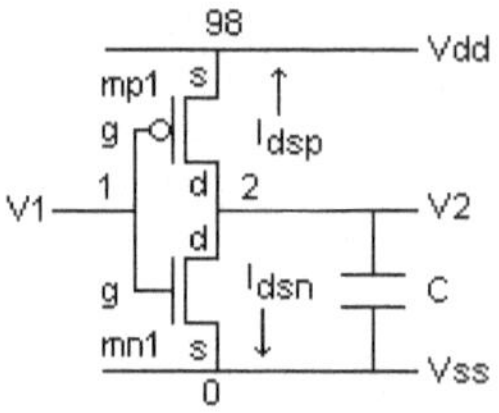

The response to a step function produces values for the delay parameters t_{PHL} and t_{PLH}. Input waveforms are approximated by linear ramps rising and falling between 0V and 1.8V in this process. To learn how t_{PHL} and t_{PLH} depend on rise and fall time, we compare output response delays to input steps and 100ps input ramps (Figure 70511). Then we compare output

Figure 705 Inverter

response delays to input steps and 400ps input ramps (Figure 70521). Spice simulations show a 20 and 45ps increase in circuit delay when switching time changes from 1ps to 100ps, and 1ps to 400ps.

> *A 100ps/400ps input change reduces to a 20ps/45ps output change.*

In Figures 70511 and 70521 the input step function is shifted in time so that the *output* signals are superimposed. Then the change in delay is the difference in 50% level (0.9V) crossings of the step and ramp inputs.

Spice Program 7051 1ps and 100ps
```
Fig7051.ckt Inverter Transient Response
.include 180_N1P1.txt
.lib G:\!B\NLPbooks\ee105pdf\cm07spice\mos018.lib
Vdd 98 0 DC 1.8
V1  1 0 PULSE(0 1.8  100p 100p 100p 300p 1000p)
V11 11 0 PULSE(0 1.75 160p  0   0 400p 1000p)
XMP1 2 1 98 98 mp9
XMN1 2 1 0   mn6
C2   2 0 90f
XMP11 12 11 98 98 mp9
XMN11 12 11 0   mn6
C12  12 0 90f
.TRAN 2e-011 1e-009 0 2e-011
.TEMP 27
.PLOT TRAN V(1) V(2) V(11) V(12) 0,2
.end
```

Problem 701 Reference Spice program 7051. Change the TRAN line to display 0 to 0.25 nanoseconds. Measure the V_1 to V_{11} delay (20ps).

Problem 702 Reference Spice program 7052. Change the TRAN line to display 0 to 0.5 nanoseconds. Measure the V_1 to V_{11} delay (45ps).

Figure 70511 Delay vs Input Switching Time – V_{11} step and V_1 100ps

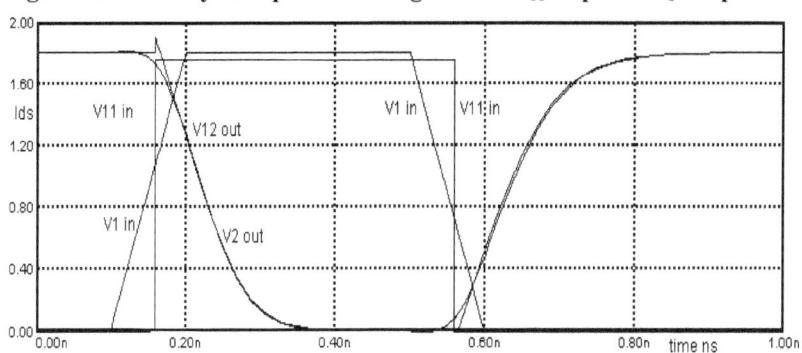

Figure 70521 Delay vs Input Switching Time - V_{11} step and V_1 400ps

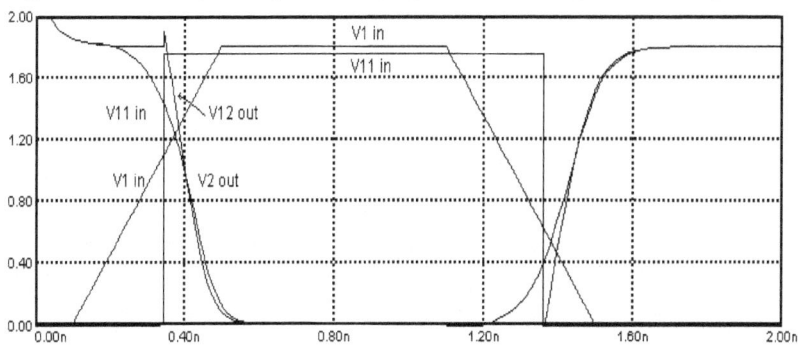

Spice Program 7052 1ps and 400ps
Fig7052.ckt Inverter Transient Response
.include 180_N1P1.txt
.lib G:\!B\NLPbooks\ee105pdf\cm07spice\mos018.lib
Vdd 98 0 DC 1.8
V1 1 0 PULSE(0 1.8 100p 400p 400p 600p 2000p)
V11 11 0 PULSE(0 1.75 345p 0   0 1015p 2000p)

XMP1 2 1 98 98 mp9
XMN1 2 1 0   mn6
C2   2 0 90f IC=1.8
XMP11 12 11 98 98 mp9
XMN11 12 11 0   mn6
C12   12 0 90f IC=1.8

.TRAN 2e-011 2e-009 0 2e-011 UIC
.TEMP 27
.PLOT TRAN V(1) V(2) V(11) V(12) 0,2
.end

7.3 Fanout, Switching time, and Clock Upper Limit

We derive what any logic system design requires − an equation for fanout and system levels' transition time Δt (L to H, H to L)

Fanout Fanout is the maximum number of circuit inputs any circuit output is designed to drive. Digital circuits usually implement a family of logic functions. If the fanout is 5, then a logic circuit output can drive 5 logic circuit inputs. A fanout of 1 is not satisfactory, because logic function requirements require circuits to drive 2, 3, 4, ... , 50 circuits. In a practical system design we need at least a fanout of 5 for most circuits, and more for special circuits.

Switching time In a CMOS system any output drives a capacitor, because the input of any CMOS circuit is a capacitor. The lines connecting circuit outputs to inputs have capacitance to ground, and some series resistance which is taken into account in Section 7.4.3 page 165. In effect the loads on circuit outputs are capacitors whose *vi* constraint equation connects switching time Δt to fanout as shown in upcoming paragraphs.

Figure 705 Inverter

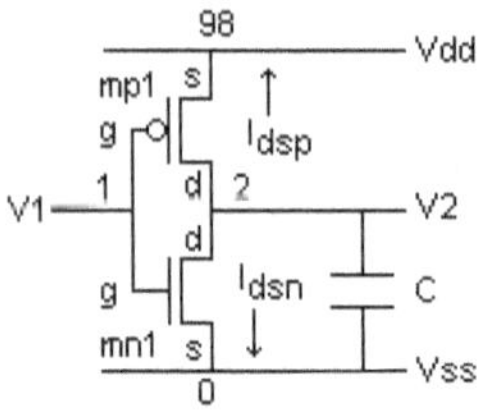

A digital signal source V_1 driving an inverter (Figure 705) switches between L (V_{ss} or 0v) and H (V_{dd} or 1.8V). If V_1 is L, then V_2 is H. When V_1 switches from L to H pmos mp_1 turns off and nmos mn_1 turns on to discharge C from H to L along path AA-BB-CC in time Δt (Figure 706)

Figure 706 Path AA-BB-CC is traversed by the nmos operating point

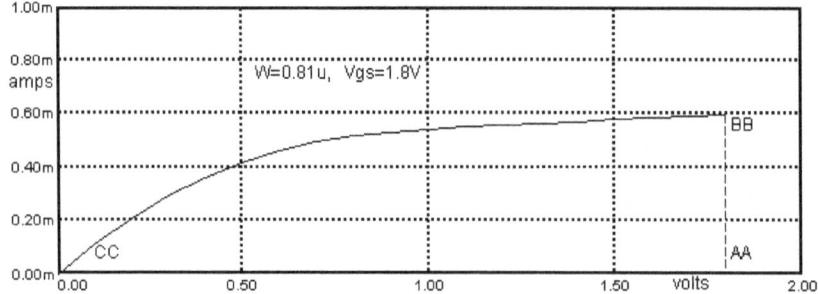

When V_1 switches from H to L nmos mn_1 turns off, and pmos mp_1 turns on. This charges load capacitor C to 1.8V in time Δt along path CC-BB-AA. This Δt is the switching time at *internal* nodes of logic circuits.

Assume an H to L digital signal source drives a tree of logic circuits where each logic circuit drives a fanout-load of F circuits. Inside an assembly of logic circuits, one circuit's output node is another circuit's input node, and so Δt becomes the logic system *internal* transition time. The output signals (Figures 70511, 70521 page 155) show that the L to H or H to L switching time of an *external input* digital signal source connected to an *external* node does not affect the system *transition* time significantly at internal nodes.

Upper clock limit   In any clock period T any node voltage will switch from H to L in time Δt, or from L to H in time Δt. In addition any node voltage must rest at H and L for time $t_{setup}+t_{hold}$ in order to satisfy setup and hold time requirements of flip-flops (sidebar The Setup and Hold Window page 90). The clock period T must be greater than $\Delta t+t_{setup}+t_{hold}$.

7.3.1 Transistor Switching Time depends on L_{min}

We compare inverter transient response with $L_{min}=2\lambda=2\mu m$ and $0.18\mu m$. C_{Load} is set equal to zero. One inverter uses the long channel $L_{min}=2\mu m$ process ($2\lambda=2\mu m$), and the other inverter uses the short channel $L_{min}=0.18\mu m$ process ($2\lambda=0.18\mu m$). The long channel Δt is 100ps (Figure 70531) and the short channel Δt is 15ps (Figure 70541).

Note the 1000ps (Figure 70531 long channel) and 100ps (Figure 70541 short channel) x-axis time scales. Also note the ratio of L_{min} is 2/0.18=11. The waveforms are almost identical, and the ratio of the time scales is 10/1. Also note that the input pulse rise is a step, and the input pulse fall is a ramp to show the responses to steps and ramps.

Figure 70531 Long Channel Inverter (2µm) switching time is about 100ps

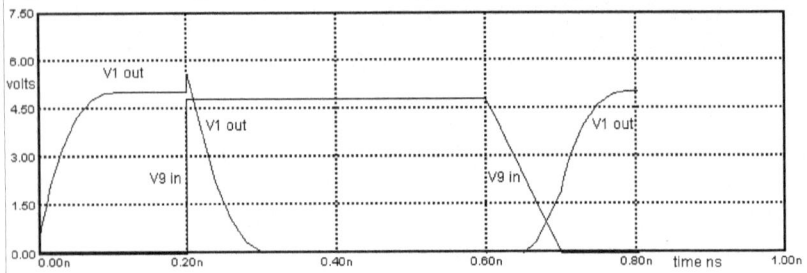

Figure 70541 Short Channel Inverter (0.18µm) switching time is about 15ps

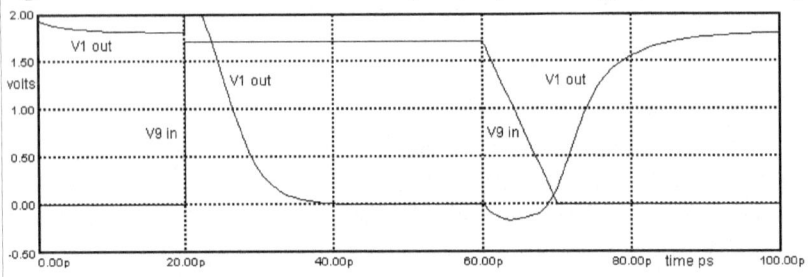

Spice Program 7053 step and ramp response
Fig7053.ckt Inverter Transient Response
.lib G:\!B\NLPbooks\ee105pdf\cm07spice\BSIM1MOD.TXT
Vdd 98 0 DC 5
V9 9 0 PULSE(0 4.8 200p 0p 100p 400p 1000p)
MP1 1 9 98 98 LongP1 L=2u W=20u ;20/2=W/L
MN1 1 9 0   0 LongN1 L=2u W=10u ;10/2
C1 1 0 1f IC=5
.TRAN 4e-011 1e-009 0 UIC
.TEMP 27
.PLOT TRAN V(9) V(1) 0,7.5
.end

Spice Program 7054 step and ramp response
Fig7054.ckt Inverter Transient Response
.include 180_N1P1.txt
Vdd 98 0 DC 1.8
V9 9 0 PULSE(0 1.7 20p 0p 10p 40p 200p)
MP1 1 9 98 98 P1 L=0.18u W=1.8u ;20/2=W/L
MN1 1 9 0 0 N1 L=0.18u W=0.9u ;10/2
C1 1 0 1f IC=1.8

.TRAN 1e-012 1e-010 0 1e-012 UIC
.TEMP 27
.PLOT TRAN V(9) V(1) -0.5,2
.end

7.3.2 Fanout Equations and System Δt

In a CMOS logic system fanout F equals the ratio of capacitors C_{Load}/C_{in}, because C_{load} equals approximately some number of C_{in}. Each capacitor can be expressed in terms of mos parameters and system switching time Δt required to traverse system rail-to-rail voltage Δv. The resulting equation for F is used to design the system.

$$(6) \quad i = C\frac{dv}{dt} \qquad C_{in} = W_{in}LC_{ox} \qquad C_L = I_L\frac{\Delta t}{\Delta V}$$

$$\boxed{(7) \quad F = \frac{C_L}{C_{in}} = \frac{1}{C_{ox}}\cdot\frac{1}{L}\cdot\frac{I_L}{W_{in}}\cdot\frac{\Delta t}{\Delta V}}$$

Here is the equation for Δt.

$$(8) \quad I_L = C_L\frac{\Delta V}{\Delta t} \qquad C_L = FC_{in} \qquad I_L = I_{ndrive}W_n \qquad C_{in} = C_{ox}L(W_n + W_p)$$

$$(9) \quad \Delta t = \frac{C_L\Delta V}{I_L} = \frac{FC_{in}\Delta V}{I_{ndrive}W_n} = \frac{FC_{ox}L(W_n + W_p)\Delta V}{I_{ndrive}W_n} = FC_{ox}\frac{L}{I_{ndrive}}\frac{\Delta V}{}\left(1 + \frac{W_p}{W_n}\right)$$

The switching time does not directly depend depend on transistor sizes, because W's enter into the equation as a ratio.

Long channel Δt In Figure 70531 the long channel *no load* Δt =100ps. A switching time estimate for an inverter *with a load F C_{in}* is as follows.

$$(10a) \quad \Delta t_{long} = F_{long}\frac{0.8\,fF}{\mu m^2}\frac{2\mu m\times 5V}{0.133mA/\mu m}(1+2) = 180F_{long}\ ps$$

$$(10b) \quad \text{If } F_{long} = 5, \text{then } \Delta t_{long} = 900\,ps$$

Compare to Figure 70552

Short channel Δt In Figure 70541 the short channel *no load* Δt =15ps. A switching time estimate for an inverter *with a load F C_{in}* is as follows.

$$(11a) \quad \Delta t_{short} = F_{short}\frac{8.63\,fF}{\mu m^2}\frac{0.18\mu m\times 1.8V}{0.61mA/\mu m}(1+2) = 13.75F_{short}\ ps$$

$$(11b) \quad \text{If } F_{short} = 5, \text{ then } \Delta t_{short} = 68.8\,ps$$

Compare to Figure 70562

Spice Program 7055

```
Fig7055.ckt  Inverter Transient Response
.lib G:\!B\NLPbooks\ee105pdf\cm07spice\BSIM1MOD.TXT
Vdd 98  0  DC 5
MP1 2 1 98 98 LongP1  L=2u  W=20u  ;20/2=W/L
MN1 2 1  0  0 LongN1  L=2u  W=10u   ; 9/2
C2  2 0 200f IC=4.9

V1  1 0 PULSE(0 4.9  0p 200p 200p 1.3n 2.5n)
*.PLOT TRAN (IG(MN1)+IG(MP1)) -1M,1M
.TRAN 1e-010 5e-009 0 1e-011 UIC
.TEMP 27
.PLOT TRAN V(1) V(2) 0,5
.end
```

Figure 70551 Long Channel Inverter gate current

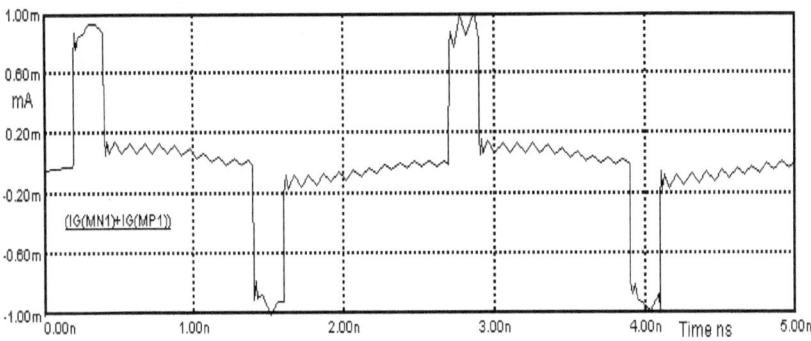

Figure 70552 Long Channel Inverter transient response

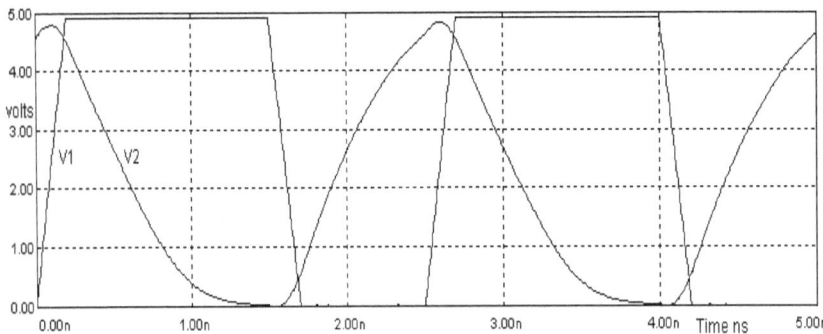

Spice Program 7056

Fig7056.ckt Inverter Transient Response
.include 180_N1P1.txt
Vdd 98 0 DC 1.8
V9 9 0 PULSE(0 1.8 0p 100p 100p 300p 1000p)

MP1 1 9 98 98 P1 L=.18u W=1.8u   ;20/2=W/L
MN1 1 9 0 0 N1 L=.18u W=0.9u   ;10/2
C1 1 0 21f   ; 5 x 4.2

.IC V(1)=5
*.PLOT TRAN (IG(MN1)+IG(MP1)) -0.25M,0.25M
.TRAN 1e-011 2e-009 0 5e-012 UIC
.TEMP 27
.PLOT TRAN V(9) V(1) 0,2
.end

Figure 70561 Short Channel Inverter gate current

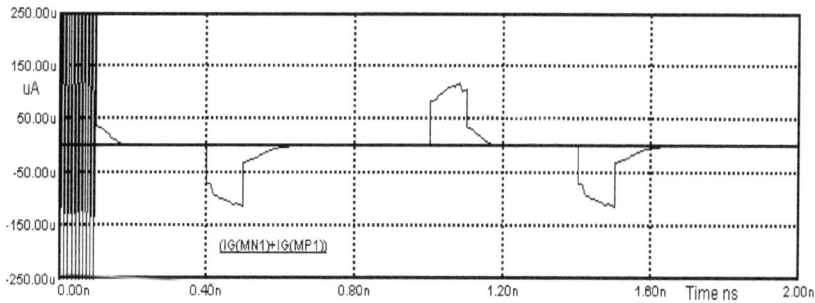

Figure 70562 Inverter transient response

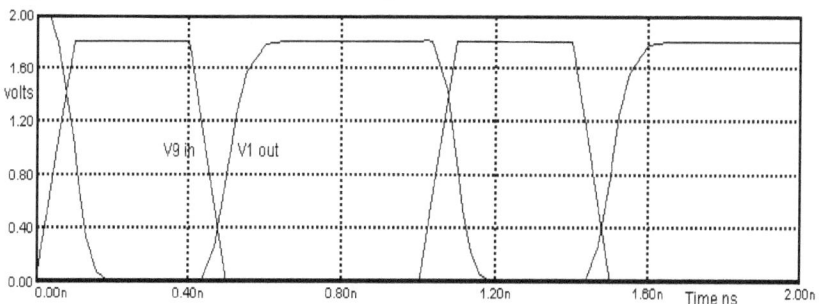

7.4 Wire Delay

Wires used to transmit signals have series resistance R, capacitance to ground C, and series inductance L. Conductance G to ground is essentially zero. In other words wires are transmission lines with RLC parameters.

The analysis method used to analyze a circuit consisting of a source driving a wire depends on the ratio of the rise time of the source signal to the travel time for a wave to traverse the length of the wire.

A lumped parameter RC approximation is satisfactory when

$$(12) \quad t_{rise} \; or \; t_{fall} > 6 \times \frac{length \; of \; line}{signal \; velocity} = 6 \times travel \; time$$

Otherwise a transmission line solution is required.

7.4.1 Resistance of Materials

Typical order-of-magnitude resistance values for an integrated circuit process are as follows.

Material	Ω/square		Material	Ω/square
p+active	2		metal1	0.07
n+active	2		metal2	0.07
poly1	7.8		metal3	0.05

Resistor Technology Ohm's law is the voltage-current constraint for resistance. The unit of resistance is the ohm (Ω). One ohm is the resistance of a material that has a one volt drop across it when one ampere flows through it.

$$(12) \quad v = iR \; \; or \; \; i = \frac{v}{R} \; \; or \; \; R = \frac{v}{i}$$

Note: In upcoming paragraphs L and W are length and width of a wire with rectangular cross-section t×W. (i.e. here L and W are **not** the mos transistor parameters.)

Resistivity In practice different dimensional units are found in handbooks. Resistivity is a function of temperature so one must always specify the temperature. For example:

(13) $At\ 20°\ C:\ \rho_{al} = \rho_{aluminum} = 2.83\mu\Omega - cm \qquad \rho_{cu} = \rho_{copper} = 1.72\mu\Omega - cm$

(14) $R(ohms) = \dfrac{resistivity \times length}{cross\ sectional\ area} = \dfrac{\rho L}{tW}$

the unit for resistivity is derived as follows :

(15) $\rho = Rt\dfrac{W}{L} \Rightarrow ohm \times meter \times \dfrac{meter}{meter} = ohm - meter$

Resistance of flat sheets (ohms per square)

Our intuition tells us that a long piece of material such as a wire has more resistance than a short piece, and as the cross section becomes larger the resistance decreases. In many applications, layers of material with uniform thickness are used. The resistance of a square of any size in these layers or 'sheets" is a constant. Here is why. The trick is to divide a piece of material into squares W units on a side. Then material of length L and width W is L/W squares in a row (16a).

(16a) $\dfrac{material\ area}{square} = \dfrac{LW}{W^2} = \dfrac{L}{W}\left(squares\ in\ a\ piece\ of\ metal\ L\ long \times W\ wide\right)$

(16b) $R(ohms) = \dfrac{resistivity \times length}{thickness \times width} = \dfrac{\rho L}{tW} = \dfrac{\rho}{t}\dfrac{ohm-meters}{meters} \times \dfrac{L}{W} squares$

(16c) $this\ implies\ the\ \dfrac{\rho}{t}\ dimensions\ are\ \dfrac{ohms}{square} \Rightarrow R = \dfrac{\rho}{t}\dfrac{ohms}{square} \times \dfrac{L}{W} squares$

The resistance of metal layers used in an integrated circuit when metal thickness t is 0.6μm are

(17a) $r_{al} = \dfrac{\rho}{t} = \dfrac{2.83}{0.6}\dfrac{10^{-6}\Omega - cm}{\mu m} \times \dfrac{10^4 \mu m}{cm} = 0.0472\dfrac{ohms}{square} at\ 20°C$

(17b) $r_{cu} = \dfrac{\rho}{t} = \dfrac{1.72}{0.6}\dfrac{10^{-6}\Omega - cm}{\mu m} \times \dfrac{10^4 \mu m}{cm} = 0.0287\dfrac{ohms}{square} at\ 20°C$

A rectangular aluminum wire 0.6 microns thick, 2 microns wide and one cm, or 10000 microns long, has 5000 squares and a total resistance of 236 ohms.

(18) $R = \dfrac{\rho}{t} \times \dfrac{L}{W} = 0.0472\dfrac{ohm}{square} \times \dfrac{10^4 \mu m}{2\mu m} squares = 0.0472 \times 5000 = 236\Omega$

A rectangular printed circuit board trace 5 mils wide and 2.7 mils thick (2oz copper) and 10cm long has 787 squares and a total resistance of 0.197 ohms.

$$(19a) \quad r_{cu} = \frac{\rho}{t} = \frac{1.72 \times 10^{-6} \Omega - cm}{2.7 \times 10^{-3} \, inch} \times \frac{1 inch}{2.54 cm} = 0.25 \times 10^{-3} \, ohms \, per \, square \, at \, 20°C$$

$$(19b) \quad \frac{L}{W} = \frac{10 cm}{5 \times 10^{-3} \, inch} \times \frac{1 inch}{2.54 cm} = 787.4 \, squares$$

$$(19c) \quad R = 787.4 \, squares \times 0.25 \times 10^{-3} \frac{ohms}{square} = 197 \times 10^{-3} \Omega$$

Temperature effect Resistivity of aluminum decreases $210 \mu\Omega$/square per °K as temperature decreases. The resistance of $0.6 \mu m$ thick aluminum wires decreases from 0.07 to 0.05 Ω/square when temperature decreases from 363°K (90°C) to 273°K (0°C). This is about a 30% decrease. Copper also decreases about 30% over this range.

7.4.2 Capacitance of Materials

Typical order-of-magnitude capacitance values for an integrated circuit process are as follows.

Capacitor	$aF / \mu m^2$	Capacitor	$aF / \mu m^2$
poly to substrate	91	metal1 to metal2	36
poly to metal1	58	metal1 to metal3	14
poly to metal2	17	metal2 to substrate	20
poly to metal3	10	metal2 to metal3	33
metal1 to substrate	42	metal3 to substrate	15

As a check on these values consider these estimates. If the field oxide between any two layers is $1 \mu m$ thick then the capacitance per square micron is 34.5aF (20b). Compare to $metal_1$/$metal_2$ and $metal_2$/$metal_3$. For example $metal_1$ to $metal_3$ capacitance is 14aF, because two oxide layers and the $metal_2$ layer separate metal layers 1 & 3.

$$(20a) \quad C_{ox} = \frac{\varepsilon_{ox}}{t_{ox}} = \frac{\varepsilon_{SiO2}\varepsilon_0}{t_{ox}} \qquad If \, t_{ox} = 1\mu m, \, and \, \varepsilon_{SiO2} = 3.9, then$$

$$(20b) \quad C_{ox} = \frac{\varepsilon_{SiO2}\varepsilon_0}{t_{ox}} = \frac{3.9 \times 8.85}{1} \frac{aF}{\mu m \times \mu m} = 34.5 \frac{aF}{\mu m^2}$$

7.4.3 Wire as an RC Circuit

Assume a metal$_1$ wire in an IC is 3λ wide and 3328λ long. This is a real example. If λ=0.09μm, then the wire dimensions are 0.270μm × 300μm. The wire cross-section is a rectangle 0.270μm wide × 0.4μm thick.

$$(21a) \quad r_{al} = \frac{\rho}{t} = \frac{2.83\times10^{-6}\Omega-cm}{0.4\mu m}\times\frac{10^4\mu m}{cm} = 0.0708\frac{ohms}{square} \; at \; 20°C$$

$$(21b) \quad R_{wire} = r_{al}\frac{L}{W} = 0.0708\frac{ohms}{square}\times\frac{3328\lambda}{3\lambda}squares = 78.54\Omega$$

$$(22) \quad C_{wire\_subs} = C_{ws}WL = 42\frac{aF}{\mu m^2}\times3328\lambda\times3\lambda\times(0.09)^2\frac{\mu m^2}{\lambda^2} = 3497aF$$

$$(23) \quad If \; R = 78.54\Omega, \; C = 3.497fF, \; then \; RC = 0.275ps$$

Figure 709 Inverter drive RC line

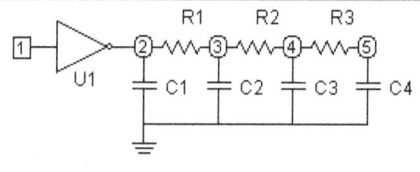

Suppose this wire drives 128 gates each with 0.85fF input capacitance. This adds 108.8fF to the 3.5fF line capacitance for a total of 112.3fF. The total line resistance is 78 ohms. The RC time constant is 8.76ps. When this loaded line is driven by an inverter and approximated as a 3 section RC line the additional delay is about 5ps.

Spice Program 7091   Wire as a 3 section RC circuit
```
Fig7091.ckt  RC Line Transient Response
.include 180_N1P1.txt
.lib G:\!B\NLPbooks\ee105pdf\cm07spice\mos018.lib
Vdd 98  0  DC 1.8
V11  1  0 PULSE(1.8 0.01  50p 100p 100p 300p 10000n)
XMP1 2 1 98 98 mp13  ;100/2
XMN1 2 1  0   mn7   ;40/2
C1 2 0 30.5f
R1 2 3 26
C2 3 0 30.5f
R2 3 4 26
C3 4 0 30.5f
R3 4 5 26
C4 5 0 30.5f

.TRAN 1e-011 1e-009 0 1e-011
.TEMP 27
.PLOT TRAN V(1) V(2) V(5) 0,2
.end
```

Figure 70911 Inverter driving RC approximation to a line, V_2 in, V_5 out

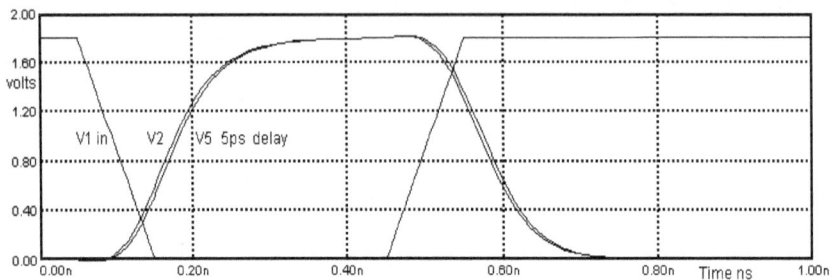

7.4.4 Wire as Transmission Line

Suppose this transmission line wire drives 128 circuits and a line with 112.3fF capacitance (7.4.3), and a 78 ohm line resistance. The result is essentially the same as the 3 RC approximation.

Figure 70921 Inverter driving Transmission line, V_2 in, V_5 out

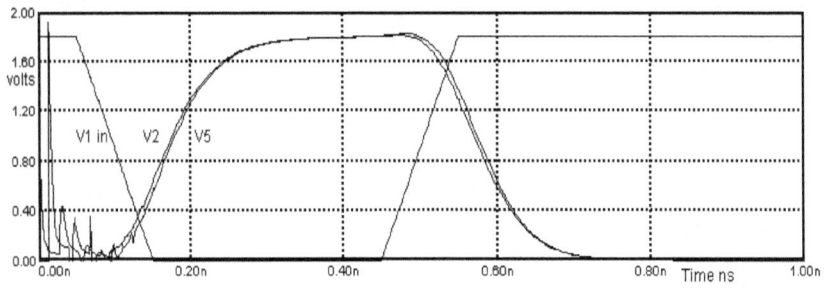

Spice Program 7092
Fig7092.ckt TLine Transient Response
.include 180_N1P1.txt
.lib G:\!B\NLPbooks\ee105pdf\cm07spice\mos018.lib
Vdd 98 0 DC 1.8
V1 1 0 PULSE(1.8 0 50p 100p 100p 300p 10000n)

XMP1 2 1 98 98 mp13   ;100/2
XMN1 2 1 0   mn7   ;40/2

T1 2 0 5 0 Len=1 C=122f R=78 L=1.5n   ; z0=100 ohms

.TRAN 1e-011 1e-009 0 2e-011 UIC
.TEMP 27
.PLOT TRAN V(1) V(2) V(5) 0,2
.end

8 How to write AC, DC, and TRAN Spice Programs

A Spice program requires a *title statement* as the first line, a (dot end) *.end statement* as the last line, and a *.temp* statement that specifies temperature.

Required program lines Between the first and last lines you insert, in any order, a list of *data statements* that describe the components of the circuit to be simulated, and a list of *control statements* that describe the circuit analysis to be performed. Any line can be empty/blank.

Comments An asterisk (*) in the first column indicates that the line is a comment line. A semicolon (;) anywhere in a line means the rest of the line is a comment. Comment lines may be placed anywhere in a Spice program.

Spice Program 2011

```
Fig2011.ckt        ;title statement must be on the first line
*The R3 line is a data statement

R3 5 7 8.2K        ;8.2K resistor connected to nodes 5 and 7
.TEMP 27
.end               ; dot end must be on last line, ends the program
```

Our numbering scheme We write a Spice program on a word processor such as Notebook or Wordpad, when we want to evaluate the performance of the circuit in Figure 201 for example. We save the text as the text file Fig2011.ckt (all Spice programs are text files). However see the note below. The first line of a program includes the name Fig2011.ckt. Plots of results from program Fig2011.ckt are labeled Fig20111, Fig20112, etc. A second program for Figure 201 is given the name Fig2012.ckt. Plots are labeled Fig20121, Fig20122, and so forth. In this way we know how circuit figures, Spice programs, Spice files, and plots are related.

> A Spice program is a text file created on any word processor.

Note: We only use the spice-text feature of the commercial Spice programs that are really complex wrap arounds to the basic Berkeley Spice. Click on File, New and select spice-text. Type into the blank screen page (your Spice may be different).

8.1 MOS *vi* Constraint

Fig6032.ckt MOS vi constraints
> *First line* The first line of every program is assumed by Spice to be a title statement. The title statement can include any words.

.include 180_N1P1.txt
> *Transistor Models* 180_N1P1.txt is a text file whose contents are shown in Appendix 2 page 184. The contents are the nmos and pmos Berkeley BSIM3V322 models for the L_{min}=0.18µm MOS fabrication process. The nmos model name is N1 and the pmos model name is P1. The dot include statement combines files by inserting the 180_N1P1.txt file contents into the Spice program.

V1 1 0 DC 0
V2 2 0 DC 0
> *Voltage sources* All sources have to be included in a Spice program. V_2 is connected to nodes 2 and 0. V_1 is connected to nodes 1 and 0. They are defined as zero volt sources. See dot DC below.

MN1 2 1 0 0 N1 L=0.18u W=1.8u ;20/2=W/L
> *Circuit components* The transistor is assigned the unique name MN_1, and uses model N1. Nodes 2 1 0 0 are the d g s b nodes (drain, gate, source, body nodes, Figure 608). Node 0 is always reserved for ground.

.DC LIN V2 0 1.8 0.05 LIN V1 0.4 1.8 0.2
.PLOT DC ID(MN1) 0,1.5M
> *Control statements* The .DC line controls the V_2 and V_1 sources .
> Dot DC sweeps V_2 from 0 to 1.8V in LIN (linear) 0.05V steps. Each sweep is executed for a fixed value of V_1 (0.4, 0.6, ... , 1.8V). I_{DS} is essentially zero when V_1=0.4V. After dot plot (.plot), and before the variables, you enter dc (without the dot). Dot plot DC executes the dot DC LIN statement. The dot print statements produce *numeric output* data.

.TEMP 27
> *Temperature* Dot temp (.temp) defines temperature as 27 degrees C.

.end
> *Last line* The last line of every program is .end (dot end).

Spice Program 6032

Fig6032.ckt MOS vi constraints
.include 180_N1P1.txt
V1 1 0 DC 0
V2 2 0 DC 0
*   2 1 0 0 are the drain gate source body nodes
MN1 2 1 0 0 N1 L=0.18u W=1.8u   ; 20/2=W/L

.DC LIN V2 0 1.8 0.05 LIN V1 0.4 1.8 0.2 ;these two lines
.PLOT DC ID(MN1) 0,1.5M   ;plot figure 60321
.TEMP 27
.end

Figure 603   **Figure 608**

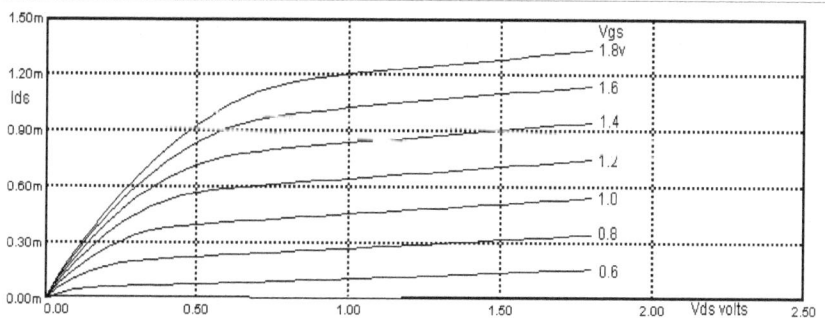

Figure 60321 nmos vi constraint

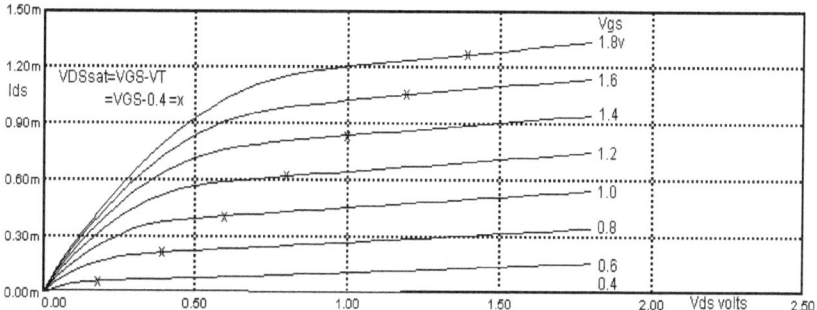

Figure 60321 nmos vi constraint showing V_{DSsat}

169

8.2 VFA MOS Operational Amplifier

Fig7072.ckt   cmos diff amp L=0.90u and Loutput=0.18u

First line The first line of every program is assumed by Spice to be a title statement. The title statement can include any words.

.include 180_N1P1.txt
.lib C:\EE103_CD\ee103spice\mos018_2.lib

Transistor Models 180_N1P1.txt is a text file whose contents are shown in Appendix 2 page 184. The contents are the nmos and pmos Berkeley BSIM3V322 models for the $L_{min}=0.18\mu m$ MOS fabrication process. The nmos model name is N1 and the pmos model name is P1. The dot include statement combines files by inserting the 180_N1P1.txt text contents into the Spice program. The text file mos018_2.lib defines transistors with specific W and L and parasitic capacitors.

V7 7 0 DC 1.8
V9 9 0 DC -1.8
V20 20 0 DC -0.0002   ; V20 is in series with v21

Voltage sources The sources are defined as 1.8V, ˜1.8V, ˜0.0002V sources. See dot DC below.

V21 21 20 DC 0 AC 1 sin(0 0.001 1e5 0 0)
Vxx +node -node   SIN(V_{offset} $V_{amplitude}$ f_o T_{delay} d)

Signal source A 1mV AC sine wave with a 100K Hertz frequency.

mn0 7 7 1 9 N1 L=1.98u W=0.18u   ;W/L=2/22 IDS=21uA
xmn4 1 1 9 9 mn7L   ;W/L=200/10
xmn5 3 1 9 9 mn7L   ;200/10

xmp1 4 4 7 7 mp3L   ;50/10
xmp2 5 4 7 7 mp3L
xmn1 4 2 3 9 mn7L   ;200/10
xmn2 5 6 3 9 mn7L

xmp3 10 5 7 7 mp3L   ; 50/10
xmn7 10 10 11 9 mn3   ; 10/2
xmn6 11 1 9 9 mn5L   ; 100/10

xmp4 7 10 8 7 mp14   ;200/2
xmn3 8 10 9 9 mn11   ; 80/2

Circuit components The MOS transistor is assigned the unique name mn_0, and uses model N1. Nodes 7 7 1 9 are the d g s b nodes (drain, gate, source, body nodes). The sub circuit xmn4 specifies an MOS transistor.

```
.AC DEC 1000 100000 1e+010
.PLOT AC VDB(5) VDB(10) VDB(8) -30,70
.PRINT AC V(5) V(10) V(8)
.PRINT AC VDB(5) VDB(10) VDB(8)

.DC V21 -0.05 0.05 0.001
.PLOT DC (10*V(21)) V(8) -2,2

.TRAN 1e-007 0.0001 0 1e-008
.PLOT TRAN (200*V(21)) V(8) -1,1
```

Control statements Dot AC, dot DC, and dot TRAN lines can coexist, because dot PLOT statements execute them. E.g. Executing a .PLOT TRAN line only executes the .TRAN line.

```
.TEMP 27
```

Temperature Dot temp (.temp) defines temperature as 27 degrees C.

```
.end
```

Last line The last line of every program is .end (dot end).

Figure 813 VFA is an Op Amp

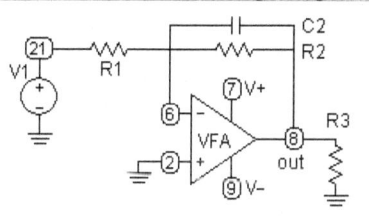

Figure 707 VFA MOS Operational Amplifier

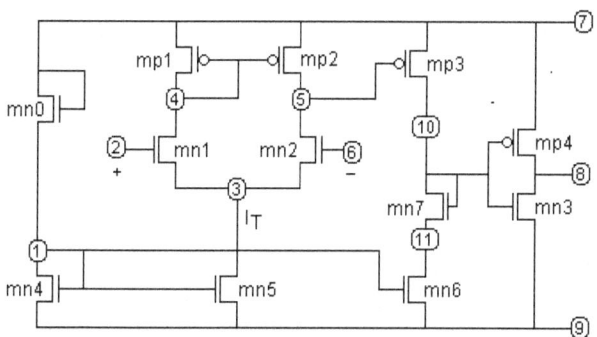

171

Spice Program 7072 Amplifier/ Mirror L=0.90μm, Output L=0.20μm

```
Fig7072.ckt      cmos diff amp L=0.90u and Loutput=0.18u
.include 180_N1P1.txt
.lib C:\EE103_CD\ee103spice\mos018_2.lib


V7 7 0 DC 1.8
V9 9 0 DC -1.8
V21 21 20 DC 0 AC 1 sin(0 0.001 1e5 0 0)
V20 20  0 DC -0.0002
R20 2 0 1k


R1 21 6 1K
R2 6 8 1000K              ;10K      ;100K       ;1000K
C2 6 8 7f                 ;220f     ;45f        ;7fF


mn0 7 7 1 9 N1 L=1.98u W=0.18u          ;W/L=2/22 IDS=21uA
xmn4 1 1 9 9 mn7L         ;W/L=200/10
xmn5 3 1 9 9 mn7L         ;200/10


xmp1 4 4 7 7 mp3L         ;50/10
xmp2 5 4 7 7 mp3L
xmn1 4 2 3 9 mn7L         ;200/10
xmn2 5 6 3 9 mn7L


xmp3 10 5 7 7 mp3L        ; 50/10
xmn7 10 10 11 9 mn3       ; 10/2
xmn6 11 1 9 9 mn5L        ; 100/10


xmp4 7 10 8 7 mp14        ;200/2
xmn3 8 10 9 9 mn11        ; 80/2

.AC DEC 1000 100000 1e+010
.PLOT AC VDB(5) VDB(10) VDB(8) -30,70
.PRINT AC V(5) V(10) V(8)
.PRINT AC VDB(5) VDB(10) VDB(8)

.DC V21 -0.05 0.05 0.001
.PLOT DC (10*V(21)) V(8) -2,2

.TRAN 1e-007 0.0001 0 1e-008
.PLOT TRAN (200*V(21)) V(8) -1,1

.TEMP 27
.end
```

gure 70721 Frequency Response Gain 1000=60dB, R₂/R₁=1000, C₂=7fF

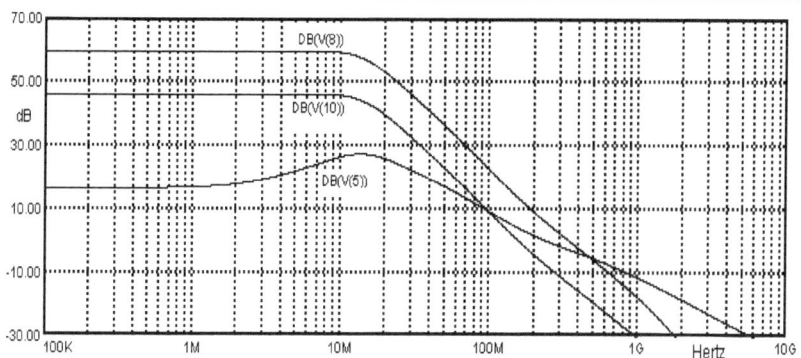

Figure 70722 MOS Op Amp DC Gain =1000=60dB, with Feedback

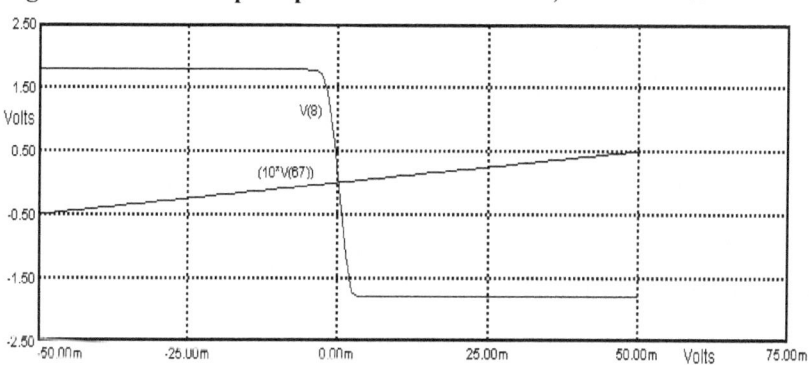

Figure 70723 MOS Op Amp Transient Response R₂/R₁=1000, C₂=7fF

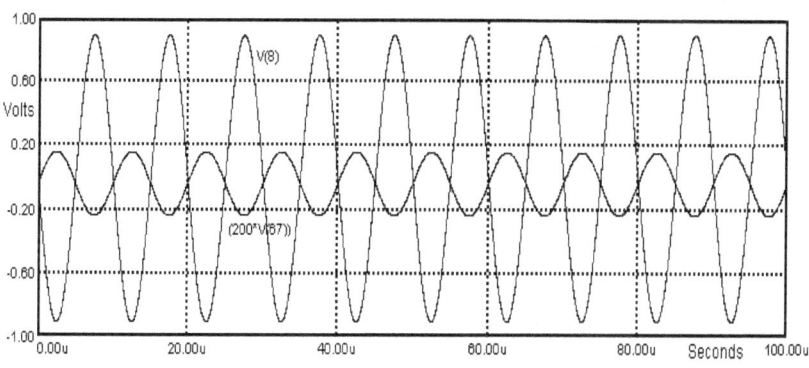

8.3 CMOS Differential Amplifier

Here we emphasize features of Spice not used in prior programs.

```
.include 180_N1P1.txt
*Now we can use models N1 and P1

.subckt mn8L 124 123 122 99
MN8L  124 123 122 99 N1  L=0.90u  W=4.50u
+ AD=2.2275p  AS=2.2275p  PD=9.99u  PS=9.99u   ;50/10=W/L
.ends mn8L

.subckt mp8L 224 223 222 298
MP8L  224 223 222 298 P1  L=0.90u  W=4.50u
+ AD=2.2275p  AS=2.2275p  PD=9.99u  PS=9.99u   ;50/10=W/L
.ends mp8L
```

```
* subckts are explained in a Section 7.2 sidebar page 113.
* sub circuit names mn8L and mp8L are transistor names.

xmp1 4 4 7 7 mp8L
xmp2 5 4 7 7 mp8L
xmn1 4 2 3 9 mn8L
xmn2 5 6 3 9 mn8L
I1   3 9 DC 200u
*subckt names begin with an x or X.
*the circuit using/calling subckts.
```

```
R2 2 0 1E10
R6 6 0 0
*mos gates are capacitors. R2, R6 provide the required DC
 paths to ground.
```

```
.PLOT AC VDB(3) VDB(4) VDB(5) -70,30
.PRINT AC V(5)
.PRINT AC VDB(3) VDB(4) VDB(5)
* dot prints produce numeric output
```

Figure 702 Differential Amplifier

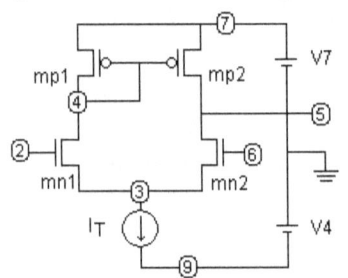

Spice Program 7024

```
Fig7024.ckt  cmos diff amp L=0.90u

.include 180_N1P1.txt

.subckt mn8L 124 123 122 99
MN8L 124 123 122 99 N1 L=0.90u W=4.50u
+ AD=2.2275p AS=2.2275p PD=9.99u PS=9.99u          ;50/10=W/L
.ends mn8L

.subckt mp8L 224 223 222 298
MP8L 224 223 222 298 P1 L=0.90u W=4.50u
+ AD=2.2275p AS=2.2275p PD=9.99u PS=9.99u          ;50/10=W/L
.ends mp8L

V7 7 0 DC  1.8
V9 9 0 DC -1.8

V2 2 0 DC 0 AC 1
R2 2 0 1E10
R6 6 0 0
xmp1 4 4 7 7 mp8L
xmp2 5 4 7 7 mp8L
xmn1 4 2 3 9 mn8L
xmn2 5 6 3 9 mn8L
I1   3 9 DC 200u
*.PLOT AC VP(3) VP(4) VP(5) -200,50
.AC DEC 1000 1e+006 1e+012
.TEMP 27
.PLOT AC VDB(3) VDB(4) VDB(5) -70,30
.PRINT AC V(5)
.PRINT AC VDB(3) VDB(4) VDB(5)
.end
```

Figure 70241 Differential Amplifier

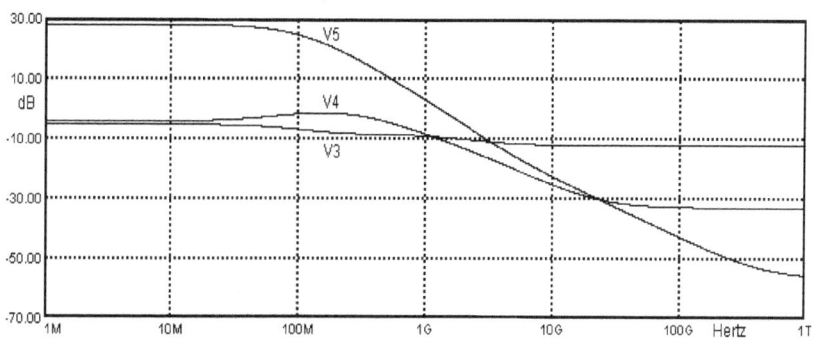

8.4 CFA MOS Operational Amplifier

Fig9091.ckt CFA Inverting op amp
> *First line* The first line of every program is assumed by Spice to be a title statement. The title statement can include any words.

```
VP  4  0  DC  5
VN  5  0  DC -5
*      PULSE(Vbase Vmax Tdelay Trise Tfall Twidth Tperiod)
V1 1 0 DC 0 AC 1 PULSE(-0.200 .200 100n .1n .1n 1000n 2500n)
```
> *Voltage sources* VP and VN power the op amp. V1 defines DC, AC, and PULSE sources.

```
.lib C:\EE103_CD\ee1spice\a_THS3001.txt
```
> The file a_THS3001.txt defines the op amp model.

```
X10      0  2  4  5  3   THS3001
R1 1 2 0.51K
R2 2 3 5.110K          ; gain=10
R3 3 0 0.510K
C3 3 0 470p
```
> *Circuit components* X10 is a sub circuit using the THS3001 op amp. The R's and C build the inverting op amp circuit Figure 909.

```
.DC V1 -5 5 0.1
.PLOT DC V(1) V(3) -7.5,7.5
.PRINT DC V(1) V(3)
.TRAN 2e-008 2.5e-006 0
.PLOT TRAN V(1) V(3) -5,5
*.PLOT AC VDB(3) -160,40
.AC DEC 10 10000 1e+011
.PLOT AC VP(3) -270,180
.PRINT AC VDB(3)
```
> *AC control statements* Dot DC, dot AC, dot TRAN define parameters to be plotted.
> *Plot the data* dot PLOT DC, AC, and TRAN execute the dot DC, AC, and TRAN control statements.

```
.TEMP 27
```
> *Temperature* Dot temp (.temp) defines temperature as 27 degrees C.

```
.end
```
> *Last line* The last line of every program is .end (dot end).

Spice Program 9091 CFA Amplifier Gain = –10

Spice 9091 evaluates the performance of the TI THS3001 op amp.

```
Fig9091.ckt  CFA Inverting op amp
VP  4  0  DC  5
VN  5  0  DC -5
*      PULSE(Vbase Vmax Tdelay Trise Tfall Twidth Tperiod)
V1 1 0 DC 0 AC 1 PULSE(-0.200 .200 100n .1n .1n 1000n 2500n)
.lib C:\EE103_CD\ee1spice\a_THS3001.txt
*.SUBCKT THS3001
* Node assignments
*           non-inverting input
*           |    inverting input
*           |    |    positive supply
*           |    |    |    negative supply
*           |    |    |    |    output
*           |    |    |    |    |
*           1    2    4    5    3
X10         0    2    4    5    3    THS3001

R1 1 2 0.51K
R2 2 3 5.110K          ; gain=10
R3 3 0 0.510K
C3 3 0 470p

.DC V1 -5 5 0.1
.PLOT DC V(1) V(3) -7.5,7.5
.PRINT DC V(1) V(3)
.TRAN 2e-008 2.5e-006 0
.PLOT TRAN V(1) V(3) -5,5
*.PLOT AC VDB(3) -160,40
.AC DEC 10 10000 1e+011
.PLOT AC VP(3) -270,180
.PRINT AC VDB(3)
.TEMP 27
.END
```

Figure 909 CFA Circuit with feedback

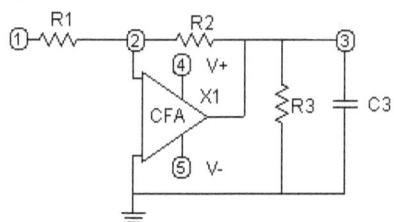

Figure 90911 CFA Op amp DC Transfer function, Gain = −10

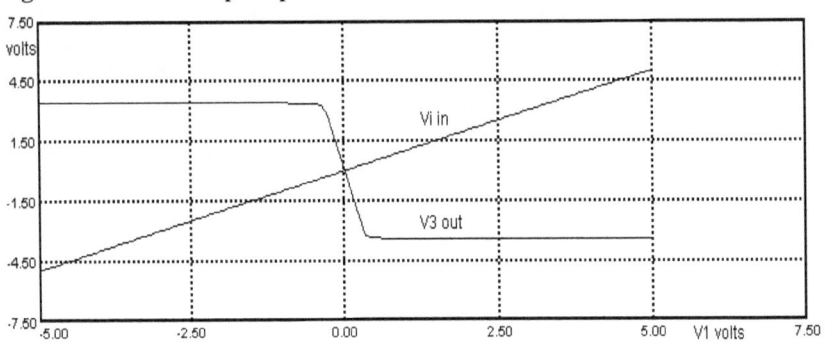

Figure 90912 CFA Op amp AC Transfer function, T(0)=20dB

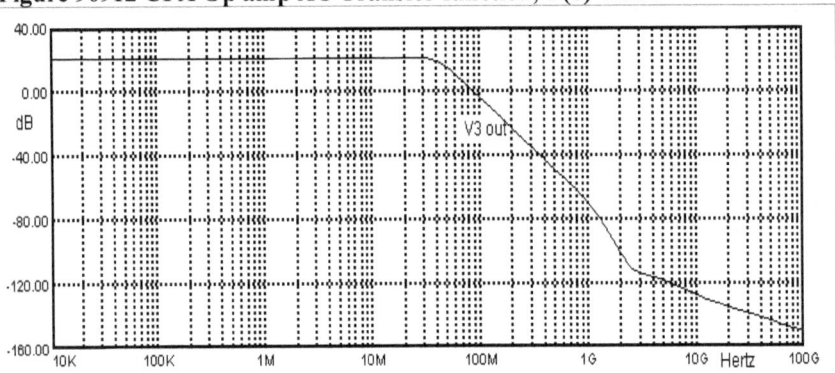

Figure 90913 CFA Op amp AC Transfer function, Phase in degrees

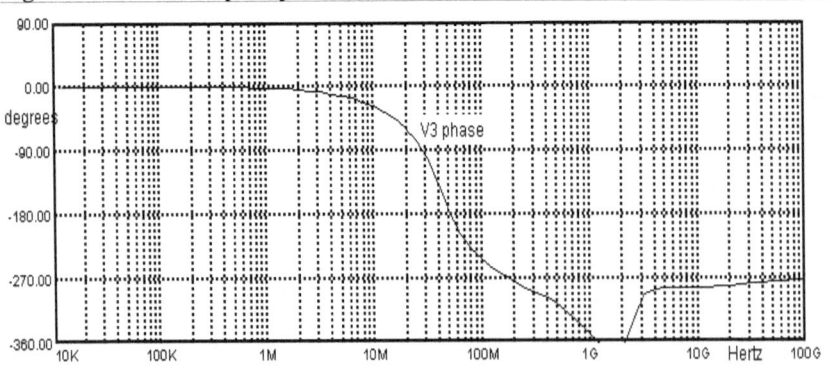

Appendix
A1 Download and install Lasi -

Procedures for downloading and installing LASI on a PC are found in upcoming paragraphs. The following assumes Lasi is NOT installed on your computer. If it is, then go to step 14.

This procedure may vary with LASI editions beyond 7.0.93, because the process seems to change with every new edition. Furthermore the instructions are in error and/or not clear. Beware!

We have done our best in what follows.

Required Equipment: PC with Windows XP or later installed.

1 *Create a folder for storing Lasi files such as C:\LasiX.*

First unwritten rule: *You have to use the C: drive.*

2 *Download the Lasi Layout Program.* The LASI files we want to download are found at *http://lasihomesite.com/*

Download *Lasi_setup_7.0.93.zip* to folder *C:\LasiX*

WARNING!! DO NOT UNZIP THESE FILES IN a folder with name C:\LASI.

3 *Extract the Lasi Layout Program in folder C:\LasiX.*

Important: LASI should always be installed using the Lasi Setup.exe program, which is now stored in *C:\LasiX.*
1) Extract folder *Lasi_setup_7.0.93* from the Zip file
2) Extract it to *C:\LasiX\ Lasi_setup_7.0.93.*

4 *Click on folder Lasi_setup_7.0.93* to get the application *Lasi_setup_7.0.93.exe*. Double click on *Lasi_setup_7.0.93.exe* to get the dialog box *Lasi_setup_7.0.93*

Everything has the same name so the process is hard to describe.

Click on *Install System*. Immediately the extracted files are stored in *C:\Lasi7*. When that finishes a new dialog box appears. Click on *Finish*. LASI is now installed in folder *C:\Lasi7*

5 *Create a project's drawing folder*

LASI is designed to be used in the context of a project folder in which all of the drawing files for that project are stored. Use any names you want.

Only STORE DRAWINGS in a drawing folder. No place else.

Open the folder *C:\LasiX\ Lasi_setup_7.0.93*.

Double click on the application *Lasi_setup_7.0.93.exe* to get the dialog box *Lasi_setup_7.0.93*

This time click on *New Drawing Folder*. Immediately a new dialog box appears. Click on *Browse*. Browse to C:\Lasi7 and click on *Open, then click on Open again.(more unwritten)* The path *C:\Lasi7* appears in the top edit box.

Note: watch out, because the instructions may NOT be accurate here. Depends on which version of Windows you have.

Add a project folder name to the *Full Existing Folder Path*, such as *C:\Lasi7\Layout_cmos_ic*.

Type an icon title such as *Layout_cmos_ic*. Click on OK to exit. The drawing folder is created. Check in the directory.

The dialog box *Lasi_setup_7.0.93* may appear. Click on *Quit*.

Note: You may get something like *"not a drawing folder."* Follow instructions to make it a drawing folder.

6 *Create a Shortcut if not done*

On the Desktop right click on the *Layout_cmos_ic* shortcut icon, select properties, and you should see the shortcut page tab.

The *Start in*: edit box should be showing *C:\Lasi7\ Layout_cmos_ic*.

If it does show that you are done – Click on *OK* to exit the shortcut.

If it does not show that click in the *Start in*: edit box to enter the box.

Backspace one character. A list box opens below the *Start in*: edit box.

ONLY use down arrow until you highlight *C:\Lasi7\ Layout_cmos_ic*. (DO NOT press Return).

Click on the highlighted line and it will appear in the "Start in" edit box. Click on OK.

Note: For the next project see step 14.

7 *Run Lasi*
On the desktop double click on the Lasi7 shortcut *Layout_cmos_ic* to start Lasi.

LASI appears on screen with the *Empty Drawing* dialog box. Click on OK.

The *Load Old Cell or Create a New Cell* dialog box appears.

Type in a cell name such as *transistor01* and click on OK. The *Create New Cell* dialog box appears. Type a *1* for new cell rank. Click on OK

Now you can work on your project. Click on *List* in the command tool bar at the top of the screen to see *transistor01* listed. Click on *Cancel*.

The folder's name *C:\Lasi7\ Layout_cmos_ic* is shown at the top of the screen. Drawings you make will be stored in the folder. Drawings are made in *cells*. A project is a set of cells.

> *Before you do anything else. This is vital:*
> Click on *Menu1* to get *Menu2*.
> Click on *Open*.
> Click on *All Layers* and click on *OK* in the dialog box.

And - ***never*** edit a Lasi file!

8 Set scale

Click on *System* (see top toolbar). Click on *Scale*
Type LAM in Physical Unit Name edit box.
Set Lasi units/Physical units to 100.
Click on OK. Read the *Important* note before you Click on *Exit*.

9 Text size

Click on Menu 1 to get Menu 2.
Click on *Tsiz*. Set Text Size to 2
Click on OK.

10 Select grids

Click on *Cnfg* (menu 2).
In the *Working Grids* and *Dot Grids* columns enter 0.1, 0.2, 0.5, 1, 2, 5, 10, 20, 50, 100.
Click on OK.

11 Define layer names and numbers

Click on *Cnfg* (menu 2).
In the editors column click on *Layers*.
Type *ARRW 1* in the *Layername Layernumber* edit box. Click on Add.
ARRW 1 appears in the Layers box above.
Repeat for each *layer/number* in the list below.
When done click on *Save*.
Click on *OK*.

ARRW 1	MET1 29	MET5 37	NSEL 46
OTLN 2	VIA1 30	VIA5 38	POL2 50
SCHM 3	MET2 31	MET6 39	CWEL 51
NTXT 4	VIA2 32	PWEL 41	PBAS 52
CTXT 5	MET3 33	NWEL 42	OVGL 53
DTXT 6	VIA3 34	ACTV 43	PADS 54
PTXT 7	MET4 35	POL1 44	
CONT 28	VIA4 36	PSEL 45	

12 Select layer colors

Click on *Attr* (menu 1). Check a box (initially ARRW 1 is checked). Enter the 3 parameters for a color. Observe the changes in the box to the left of the check box. Check the next box and enter parameter, etc. When done with all layers click on *OK*

Warning!! DO NOT Click on the "Solid", "No Fill", "No Dash" buttons.

The 4 items in each box are   Layer   Color   Fill   Dash

1 r 0 0	30 v 13 0	41 1 0 2 (letter L)	50 o 6 0
2 o 0 0	31 v 10 0	42 b 0 2	51 c 2 4
3 o 0 0	32 o 13 0	43 g 3 0	52 y 0 0
4 o 0 0	33 o 11 0	44 r 8 0	53 1 5 0
5 g 0 0	34 g 13 0	45 y 0 0	54 y 0 0
6 t 0 0	35 g 11 0	46 g 0 0	
7 v 0 0 (letter v)	36 y 13 0		
	37 y 10 0		
28 b 13 0	38 o 13 0		
29 b 11 0	39 o 11 0		

13 *Finally - Create and edit a cell*
Click on the *Load* command. This how you create new drawing cells.

14. Note: For the next project repeat step 5 and add Layout 02 or whatever name you want. Then do step 6 again.

A2 The files 180_N1P1.txt, mos018_2.lib, and mos018_L.lib

L=0.18µm BSIM3V322 Model Card (the file 180_N1P1.txt)

```
* Predictive Technology Model Beta Version
* 0.18um NMOS SPICE Parametersv (normal one)
* http://www.device.EECS.Berkeley.EDU/ ~ptm/
.model N1 NMOS        Level = 8
+Lint = 4.e-08          Tox = 4.e-09      Vth0 = 0.3999        Rdsw = 250
+lmin=1.8e-7            lmax=1.8e-7       wmin=1.8e-7          wmax=1.0e-4
+Tref=27.0              version =3.1
+Xj= 6.0000000E-08      Nch= 5.9500000E+17
+lln= 1.0000000         lwn= 1.0000000                wln= 0.00
+wwn= 0.00              ll= 0.00
+lw= 0.00               lwl= 0.00                     wint= 0.00
+wl= 0.00               ww= 0.00                      wwl= 0.00
+Mobmod=1               binunit= 2                    xl= 0
+xw=0                   binflag=0
+Dwg= 0.00              Dwb= 0.00
+K1= 0.5613000          K2= 1.0000000E-02
+K3= 0.00               Dvt0= 8.0000000               Dvt1= 0.7500000
+Dvt2= 8.0000000E-03    Dvt0w= 0.00                   Dvt1w= 0.00
+Dvt2w= 0.00            Nlx= 1.6500000E-07            W0= 0.00
+K3b= 0.00              Ngate= 5.0000000E+20
+Vsat= 1.3800000E+05    Ua= -7.0000000E-10            Ub= 3.5000000E-18
+Uc= -5.2500000E-11     Prwb= 0.00
+Prwg= 0.00             Wr= 1.0000000                 U0= 3.5000000E-02
+A0= 1.1000000          Keta= 4.0000000E-02           A1= 0.00
+A2= 1.0000000          Ags= -1.0000000E-02           B0= 0.00        B1= 0.00
+Voff= -0.12350000      NFactor= 0.9000000            Cit= 0.00
+Cdsc= 0.00             Cdscb= 0.00                   Cdscd= 0.00
+Eta0= 0.2200000        Etab= 0.00                    Dsub= 0.8000000
+Pclm= 5.0000000E-02    Pdibl1= 1.2000000E-02         Pdiblc2= 7.5000000E-03
+Pdiblcb=-1.3500000E-02 Drout= 1.7999999E-02          Pscbe1= 8.6600000E+08
+Pscbe2= 1.0000000E-20  Pvag= -0.2800000              Delta= 1.0000000E-02
+Alpha0= 0.00           Beta0= 30.0000000
+kt1= -0.3700000        kt2= -4.0000000E-02           At= 5.5000000E+04
+Ute= -1.4800000        Ua1= 9.5829000E-10            Ub1= -3.3473000E-19
+Uc1= 0.00              Kt1l= 4.0000000E-09           Prt= 0.00
+Cj= 0.00365            Mj= 0.54                      Pb= 0.982
+Cjsw= 7.9E-10          Mjsw= 0.31                    Php= 0.841
+Cta= 0                 Ctp= 0                        Pta= 0
+Ptp= 0                 JS=1.50E-08                   JSW=2.50E-13
+N=1.0                  Xti=3.0                       Cgdo=2.786E-10
+Cgso=2.786E-10           Cgbo=0.0E+00                  Capmod= 2
+NQSMOD= 0              Elm= 5                        Xpart= 1
+Cgsl= 1.6E-10          Cgdl= 1.6E-10                 Ckappa= 2.886
+Cf= 1.069e-10          Clc= 0.0000001                Cle= 0.6
+Dlc= 4E-08             Dwc= 0                        Vfbcv= -1
```

L=0.18μm BSIM3V322 Model Card (the file 180_N1P1.txt)

```
* Predictive Technology Model Beta Version
* 0.18um PMOS SPICE Parametersv (normal one)
.model P1  PMOS      Level = 8
+Lint = 3.e-08          Tox = 4.2e-09        Vth0 = -0.42      Rdsw = 450
+lmin=1.8e-7            lmax=1.8e-7          wmin=1.8e-7       wmax=1.0e-4
+Tref=27.0              version =3.1
+Xj= 7.0000000E-08      Nch= 5.9200000E+17
+lln= 1.0000000         lwn= 1.0000000                wln= 0.00
+wwn= 0.00              ll= 0.00
+lw= 0.00               lwl= 0.00            wint= 0.00
+wl= 0.00               ww= 0.00             wwl= 0.00
+Mobmod=1               binunit= 2           xl= 0.00
+xw= 0.00               binflag= 0
+Dwg= 0.00              Dwb= 0.00
+ACM= 0                 ldif=0.00            rsh= 0
+rd= 0                  rs= 0                +rsc= 0      rdc= 0
+K1= 0.5560000          K2= 0.00
+K3= 0.00               Dvt0= 11.2000000     Dvt1= 0.7200000
+Dvt2= -1.0000000E-02   Dvt0w= 0.00          Dvt1w= 0.00
+Dvt2w= 0.00            Nlx= 9.5000000E-08   W0= 0.00
+K3b= 0.00              Ngate= 5.0000000E+20
+Vsat= 1.0500000E+05    Ua= -1.2000000E-10   Ub= 1.0000000E-18
+Uc= -2.9999999E-11     Prwb= 0.00
+Prwg= 0.00             Wr= 1.0000000        U0= 8.0000000E-03
+A0= 2.1199999          Keta= 2.9999999E-02  A1= 0.00
+A2= 0.4000000          Ags= -0.1000000      B0= 0.00      B1= 0.00
+Voff= -6.40000000E-02  NFactor= 1.4000000   Cit= 0.00
+Cdsc= 0.00             Cdscb= 0.00          Cdscd= 0.00
+Eta0= 8.5000000        Etab= 0.00           Dsub= 2.8000000
+Pclm= 2.0000000        Pdiblc1= 0.1200000   Pdiblc2= 8.0000000E-05
+Pdiblcb= 0.1450000     Drout= 5.0000000E-02 Pscbe1= 1.0000000E-20
+Pscbe2=1.0000000E-20   Pvag= -6.0000000E-02 Delta= 1.0000000E-02
+Alpha0= 0.00           Beta0= 30.0000000
+kt1= -0.3700000        kt2= -4.0000000E-02  At= 5.5000000E+04
+Ute= -1.4800000        Ua1= 9.5829000E-10   Ub1= -3.3473000E-19
+Uc1= 0.00              Kt1l= 4.0000000E-09  Prt= 0.00
+Cj= 0.00138            Mj= 1.05             Pb= 1.24
+Cjsw= 1.44E-09         Mjsw= 0.43           Php= 0.841
+Cta= 0.00093           Ctp= 0               Pta= 0.00153
+Ptp= 0                 JS=1.50E-08          JSW=2.50E-13
+N=1.0                  Xti=3.0              Cgdo=2.786E-10
+Cgso=2.786E-10            Cgbo=0.0E+00         Capmod= 2
+NQSMOD= 0              Elm= 5               Xpart= 1
+Cgsl= 1.6E-10          Cgdl= 1.6E-10        Ckappa= 2.886
+Cf= 1.058e-10          Clc= 0.0000001       Cle= 0.6
+Dlc= 3E-08             Dwc= 0               Vfbcv= -1
```

*Mos018_2.lib with node 99   Spice Library   2lambda = 0.18u

*NMOS
.subckt mn1a 103 102 101 99
MN1a 103 102 101 99 N1 L=0.18u W=0.27u
+ AD=0.1337p AS=0.1337p PD=1.53u PS=1.53u ; 3/2=W/L
.ends mn1a

.subckt mn1 103 102 101 99
MN1 103 102 101 99 N1 L=0.18u W=0.36u
+ AD=0.1782p AS=0.1782p PD=1.71u PS=1.71u ; 4/2=W/L
.ends mn1

.subckt mn2 106 105 104 99
MN2 106 105 104 99 N1 L=0.18u W=0.54u
+ AD=0.2673p AS=0.2673p PD=2.07u PS=2.07u ;6/2=W/L
.ends mn2

.subckt mn3 109 108 107 99
MN3 109 108 107 99 N1 L=0.18u W=0.90u
+ AD=0.4455p AS=0.4455p PD=2.79u PS=2.79u ;10/2=W/L
.ends mn3

.subckt mn4 112 111 110 99
MN4 112 111 110 99 N1 L=0.18u W= 1.35u
+ AD=0.6683p AS=0.6683p PD=3.69u PS=3.69u ;15/2=W/L
.ends mn4

.subckt mn5 115 114 113 99
MN5 115 114 113 99 N1 L=0.18u W=1.80u
+ AD=0.8910p AS=0.8910p PD=4.59u PS=4.59u ;20/2=W/L
.ends mn5

.subckt mn6 118 117 116 99
MN6 118 117 116 99 N1 L=0.18u W=2.70u
+ AD=1.3365p AS=1.3365p PD=6.39u PS=6.39u ;30/2=W/L
.ends mn6

.subckt mn7 121 120 119 99
MN7 121 120 119 99 N1 L=0.18u W=3.60u
+ AD=1.7820p AS=1.7820p PD=8.19u PS=8.19u ;40/2=W/L
.ends mn7

.subckt mn8 124 123 122 99
MN8 124 123 122 99 N1 L=0.18u W=4.50u
+ AD=2.2275p AS=2.2275p PD=9.99u PS=9.99u ;50/2=W/L
.ends mn8

```
.subckt mn9 127 126 125 99
MN9  127 126 125 99 N1  L=0.18u  W=5.40u
+ AD=2.6730p  AS=2.6730p  PD=11.79u  PS=11.79u  ;60/2=W/L
.ends mn9

.subckt mn10 130 129 128 99
MN10 130 129 128 99 N1  L=0.18u  W=6.30u
+ AD=3.1185p  AS=3.1185p  PD=13.59u  PS=13.59u  ;70/2=W/L
.ends mn10

.subckt mn11 133 132 131 99
MN11  133 132 131 99 N1  L=0.18u  W= 7.20u
+ AD=3.5640p  AS=3.5640p  PD=15.39u  PS=15.39u  ;80/2=W/L
.ends mn11

.subckt mn12 134 133 132 99
MN12  134 133 132 99 N1  L=0.18u  W= 8.10u
+ AD=4.0095p  AS=4.0095p  PD=17.19u  PS=17.19u  ;90/2=W/L
.ends mn12

.subckt mn13 137 136 135 99
MN13  137 136 135 99 N1  L=0.18u  W=9.00u
+ AD=4.4550p  AS=4.4550p  PD=18.99u  PS=18.99u  ;100/2=W/L
.ends mn13

.subckt mn14 3 2 1 99
MN14  3  2  1  99 N1  L=0.18u  W=18.00u
+ AD=8.910p  AS=8.910p  PD=36.99u  PS=36.99u  ;200/2=W/L
.ends mn14

.subckt mn15 3  2  1 99
MN15  3  2  1  99 N1  L=0.18u  W=36.00u
+ AD=17.820p  AS=17.820p  PD=72.99u  PS=72.99u  ;400/2=W/L
.ends mn15
```

*PMOS

.subckt mp1a 203 202 201 298
MP1a 203 202 201 298 P1 L=0.18u W=0.27u
+ AD=0.1337p AS=0.1337p PD=1.53u PS=1.53u ; 3/2=W/L
.ends mp1a

.subckt mp1 203 202 201 298
MP1 203 202 201 298 P1 L=0.18u W=0.36u
+ AD=0.1782p AS=0.1782p PD=1.71u PS=1.71u ; 4/2=W/L
.ends mp1

.subckt mp2 206 205 204 298
MP2 206 205 204 298 P1 L=0.18u W=0.54u
+ AD=0.2673p AS=0.2673p PD=2.07u PS=2.07u ; 6/2=W/L
.ends mp2

.subckt mp3 209 208 207 298
MP3 209 208 207 298 P1 L=0.18u W=0.90u
+ AD=0.4455p AS=0.4455p PD=2.79u PS=2.79u ; 10/2=W/L
.ends mp3

.subckt mp4 212 211 210 298
MP4 212 211 210 298 P1 L=0.18u W= 1.35u
+ AD=0.6683p AS=0.6683p PD=3.69u PS=3.69u   ; 15/2=W/L
.ends mp4

.subckt mp5 215 214 213 298
MP5 215 214 213 298 P1 L=0.18u W=1.80u
+ AD=0.8910p AS=0.8910p PD=4.59u PS=4.59u   ; 20/2=W/L
.ends mp5

.subckt mp6 218 217 216 298
MP6 218 217 216 298 P1 L=0.18u W=2.70u
+ AD=1.3365p AS=1.3365p PD=6.39u PS=6.39u ; 30/2=W/L
.ends mp6

.subckt mp7 221 220 219 298
MP7 221 220 219 298 P1 L=0.18u W=3.60u
+ AD=1.7820p AS=1.7820p PD=8.19u PS=8.19u ; 40/2=W/L
.ends mp7

.subckt mp8 224 223 222 298
MP8 224 223 222 298 P1 L=0.18u W=4.50u
+ AD=2.2275p AS=2.2275p PD=9.99u PS=9.99u   ; 50/2=W/L
.ends mp8

```
.subckt mp9 227 226 225 298
MP9  227 226 225 298 P1  L=0.18u  W=5.40u
+ AD=2.6730p  AS=2.6730p  PD=11.79u  PS=11.79u    ; 60/2=W/L
.ends mp9

.subckt mp10 230 229 228 298
MP10 230 229 228 298 P1  L=0.18u  W=6.30u
+ AD=3.1185p  AS=3.1185p  PD=13.59u  PS=13.59u    ; 70/2=W/L
.ends mp10

.subckt mp11 233 232 231 298
MP11  233 232 231 298 P1  L=0.18u  W= 7.20u
+ AD=3.5640p  AS=3.5640p  PD=15.39u  PS=15.39u    ; 80/2=W/L
.ends mp11

.subckt mp12 234 233 232 298
MP12  234 233 232 298 P1  L=0.18u  W= 8.10u
+ AD=4.0095p AS=4.0095p  PD=17.19u  PS=17.19u    ; 90/2=W/L
.ends mp12

.subckt mp13 237 236 235 298
MP13  237 236 235 298 P1  L=0.18u  W=9.00u
+ AD=4.4550p  AS=4.4550p  PD=18.99u  PS=18.99u    ; 100/2=W/L
.ends mp13

.subckt mp14 3 2 1 298
MP14  3  2  1  298 P1  L=0.18u  W=18.00u
+ AD=8.910p  AS=8.910p  PD=36.99u  PS=36.99u      ; 200/2=W/L
.ends mp14

.subckt mp15 3  2  1 298
MP15  3  2  1  298 P1  L=0.18u  W=36.00u
+ AD=17.820p  AS=17.820p  PD=72.99u  PS=72.99u    ; 400/2=W/L
.ends mp15
```

CMOS Circuit Design

*NMOS

.subckt mn1La 103 102 101 99
MN1La 103 102 101 99 N1 L=0.90u W=1.35u
+ AD=0.6683p AS=0.6683p PD=3.690u PS=3.690u
* 15/10=W/L
.ends mn1La

.subckt mn1L 103 102 101 99
MN1L 103 102 101 99 N1 L=0.90u W=1.80u
+ AD=0.8910p AS=0.8910p PD=4.590u PS=4.590u
* 20/10=W/L
.ends mn1L

.subckt mn2L 106 105 104 99
MN2L 106 105 104 99 N1 L=0.90u W=2.70u
+ AD=1.3365p AS=1.3365p PD=6.390u PS=6.390u
*30/10=W/L
.ends mn2L

.subckt mn3L 109 108 107 99
MN3L 109 108 107 99 N1 L=0.90u W=4.50u
+ AD=2.2275p AS=2.2275p PD=9.990u PS=9.990u
*50/10=W/L
.ends mn3L

.subckt mn4L 112 111 110 99
MN4L 112 111 110 99 N1 L=0.90u W= 6.75u
+ AD=3.3413p AS=3.3413p PD=14.490u PS=14.490u
*75/10=W/L
.ends mn4L

.subckt mn5L 115 114 113 99
MN5L 115 114 113 99 N1 L=0.90u W=9.00u
+ AD=4.455p AS=4.455p PD=18.990u PS=18.990u
*100/10=W/L
.ends mn5L

.subckt mn6L 118 117 116 99
MN6L 118 117 116 99 N1 L=0.90u W=13.5u
+ AD=6.6825p AS=6.6825p PD=27.990u PS=27.990u
*150/10=W/L
.ends mn6L

```
.subckt mn7L 121 120 119 99
MN7L  121 120 119 99 N1  L=0.90u  W=18.0u
+ AD=8.910p  AS=8.910p  PD=36.990u  PS=36.990u
*200/10=W/L
.ends mn7L

.subckt mn8L 124 123 122 99
MN8L  124 123 122 99 N1  L=0.90u  W=22.5u
+ AD=11.1375p  AS=11.1375p  PD=45.990u  PS=45.990u
*250/10=W/L
.ends mn8L

.subckt mn9L 127 126 125 99
MN9L  127 126 125 99 N1  L=0.90u  W=27.0u
+ AD=13.3650p  AS=13.3650p  PD=54.990u  PS=54.990u
*300/10=W/L
.ends mn9L
```

*PMOS

```
.subckt mp1La 203 202 201 298
MP1La  203 202 201 298 P1  L=0.90u  W=1.35u
+ AD=0.6683p  AS=0.6683p  PD=3.690u  PS=3.690u
* 15/10=W/L
.ends mp1La

.subckt mp1L 203 202 201 298
MP1L  203 202 201 298 P1  L=0.90u  W=1.80u
+ AD=0.8910p  AS=0.8910p  PD=4.590u  PS=4.590u
* 20/10−W/L
.ends mp1L

.subckt mp2L 206 205 204 298
MP2L  206 205 204 298 P1  L=0.90u  W=2.70u
+ AD=1.3365p  AS=1.3365p  PD=6.390u  PS=6.390u
*30/10=W/L
.ends mp2L

.subckt mp3L 209 208 207 298
MP3L  209 208 207 298 P1  L=0.90u  W=4.50u
+ AD=2.2275p  AS=2.2275p  PD=9.990u  PS=9.990u
*50/10=W/L
.ends mp3L

.subckt mp4L 212 211 210 298
MP4L  212 211 210 298 P1  L=0.90u  W= 6.75u
+ AD=3.3413p  AS=3.3413p  PD=14.490u  PS=14.490u
*75/10=W/L
.ends mp4L
```

```
.subckt mp5L 215 214 213 298
MP5L  215 214 213 298 P1  L=0.90u  W=9.00u
+ AD=4.455p  AS=4.455p  PD=18.990u  PS=18.990u
*100/10=W/L
.ends mp5L

.subckt mp6L 218 217 216 298
MP6L  218 217 216 298 P1  L=0.90u  W=13.5u
+ AD=6.6825p  AS=6.6825p  PD=27.990u  PS=27.990u
*150/10=W/L
.ends mp6L

.subckt mp7L 221 220 219 298
MP7L  221 220 219 298 P1  L=0.90u  W=18.0u
+ AD=8.910p  AS=8.910p  PD=36.990u  PS=36.990u
*200/10=W/L
.ends mp7L

.subckt mp8L 224 223 222 298
MP8L  224 223 222 298 P1  L=0.90u  W=22.5u
+ AD=11.1375p  AS=11.1375p  PD=45.990u  PS=45.990u
*250/10=W/L
.ends mp8L

.subckt mp9L 227 226 225 298
MP9L  227 226 225 298 P1  L=0.90u  W=27.0u
+ AD=13.3650p  AS=13.3650p  PD=54.990u  PS=54.990u
*300/10=W/L
.ends mp9L
```

INDEX

www.ingramcontent.com/pod-product-compliance
Lightning Source LLC
Chambersburg PA
CBHW051456170526
45166CB00001B/268